Keene
Tahihapa
Greenwich
Camerons
Sand Creek
N
El Paso City
Dry Salt Lake
Gray's Sta.
Mojave
Yucca
Dry
Lake
Kramer
Pilot Butte
Granite Spring
Grant
Harper
Hinckley
Wellman
Calico
Sulphur Springs
Grapevine
SOUTHERN
Cottonwood
Mohave
Government Sta.
River
FA
Cam
SAN
man's Sta.
Lake
PACIFIC
Lancaster
Point of Rock Sta.
Alpine
Lang
Newhall
Andrews
Soledad City
SAN BERNARDINO RANGE
Lanes
Coyote
Huntington
SAN BER
veda
San Fernando
San Gabriel
Lexington
Citrus
Martins
Baindstown
geles
San
Spadra
Pomona
Cucamonga
San Bernardino
rence
pton
Puente
Corvallis
Norwalk
San Salvador
Ripes
Mound Sta.

Wishing you a merry & bright
2021 holiday season!
Hoping you enjoy this
memorable look at
Ventura County.
Thank you for always
remembering me when
you think of real estate!

Hugs to All, *Cindy*

LIVING LEGACY

The Story of Ventura County Agriculture

by

John Krist

With photos by

Gary Faye

Published by the Farm Bureau of Ventura County

MUSEUM of VENTURA COUNTY

With major support provided by the University of California Hansen Trust
in cooperation with the Museum of Ventura County

Text Copyright © 2007 by John Krist

Photos Copyright © 2007 by Gary Faye, except as noted

Design by Barbara Jefferies and Claudia Laub

ISBN 978-0-9798780-0-8

Published by the Farm Bureau of Ventura County
5156 McGrath St., Suite 102
Ventura, CA 93003

Library of Congress Cataloging-in-Publication Data

Krist, John, 1958-
Living Legacy: The Story of Ventura County Agriculture
128 p. 9 x 13 in.
Includes bibliographic references and index

1. Agriculture—California	2. Ventura County—California—History
3. Strawberries—California	4. Lemons—California 5. Avocados—California
6. Cattle Ranching—California	7. Land Use—California 8. Urban Sprawl—California
9. Organic Farming—California	10. Pesticides—California 11. Globalization

Faulkner House, Santa Paula.　　　　　　　　　　　　　*Gary Faye*

Contents

Preface and Acknowledgments

This book describes a year in the fields, orchards, packinghouses and nurseries of Ventura County, a slice of coastal Southern California that remains improbably dominated by farms despite its proximity to the largest metropolitan center in America.

With its balmy weather, beaches and mountains, Ventura County is an extremely desirable place to live and has the stratospheric housing prices to prove it. Yet unlike Los Angeles and Orange counties, its formerly agricultural neighbors to the south, Ventura County has so far resisted the urge to pave every acre of its rich soil to accommodate the apparently infinite number of people who would like to live here. In 2007, it contains as many acres of irrigated cropland as it does acres of housing tracts, shopping malls and industrial parks; each of its relatively compact cities is separated from the others by broad swaths of working farms and ranches.

That this blend of agriculture and suburbia persists in the 21st century, on the edge of a vast urbanized region containing 18 million people, is remarkable but not accidental. It is the result of deliberate local land-use policies designed to manage and contain urban growth, as well as the creativity and adaptability of growers who have managed to stay in business despite high operating costs and lower-priced competition from parts of the world where wages, regulatory compliance, land and water cost much less. Ventura County's varied and scenic landscape, like the farming industry that plays such a critical role in maintaining it, is a priceless legacy bequeathed to the present by thoughtful, creative and forward-thinking men and women of the past.

Living Legacy is based on a series of stories published in 2006 under the title "Farming on the Edge" by the *Ventura County Star*, the daily newspaper where I have worked since 1983. I have adapted them for this book, but the purpose remains the same: to provide a behind-the-scenes look at farming in Southern California's last great agricultural landscape, exploring the characteristics that have enabled it to prosper for more than a century even as farming in other parts of the region has succumbed to urbanization, globalization and other implacable forces. They've been paired here with a collection of remarkable images by Gary Faye, a professional photographer with a keen eye for the visual beauty of both the agricultural landscape and the crops it produces.

I chose to focus on the farmers and ranchers highlighted in these pages because they display some of the key characteristics of the county's signature industry. They include members of a fourth-generation Ventura County farming family who produce lemons and avocados, an immigrant farm laborer's son who became a strawberry grower, a rancher who runs cows and calves on rangeland once grazed by the cattle herds of an 18th century Spanish mission, and a tangerine grower who's been known to dress up like a carrot to teach schoolchildren about the glories of fresh produce.

Their stories are augmented by those of other participants in the farming business: farm laborers, cowboys, labor contractors, packinghouse managers, pilots, dockworkers and biotechnology researchers. *Living Legacy* is intended to paint a portrait of Ventura County agriculture at a pivotal moment in its long history— a moment when, despite a heritage that reaches more than 200 years into the past, the industry's future as a robust mainstay of the local economy seems in doubt.

The farmers and ranchers I profiled and followed during the year it took to research this book agreed to subject themselves to a constant barrage of intrusive questions about their lives, their businesses, their success and their failures. They were unfailingly polite, indulgent and candid, and I simply could not have done any of this without their cooperation. So, to Richard Atmore, Lisa Brenneis, Jim Churchill, Link Leavens, Leslie Leavens-Crowe, Cecil Martinez and David Schwabauer, my heartiest thanks.

In addition, I received invaluable help and cooperation from Rex Laird, chief executive officer of the Farm Bureau of Ventura County, who answered questions, shared his perspective, helped me get in touch with many of my sources, and then persuaded his organization to take on the role of book publisher when earlier arrangements fell through. Earl McPhail, the county agricultural commissioner, opened his office's archives to me, allowing me to gain a detailed appreciation of the changes in cropping patterns over the years. Tim Schiffer and Charles Johnson at the Museum of Ventura County opened their institution's priceless photo collection to me, and provided digital copies of evocative imagery from the county's agricultural past.

Edgar Terry provided countless helpful comments on the business of farming, and let me tromp around in his pepper and celery fields. Phil McGrath welcomed me into his organic farm and protected me from angry geese. Harold Edwards and Alex Teague at Limoneira shared their knowledge and let me roam around in the historical treasure that is their packinghouse.

Thanks are also due the folks at Saticoy Lemon Association, Villa Park Orchards Association, Calavo, Seminis, Saticoy Foods, Brokaw Nursery, Lassen Canyon Nursery and Associates Insectary for letting me inside their packinghouses, greenhouses and bughouses. Rick Throckmorton at Aspen Helicopters took me flying, and Will Berg at the Port of Hueneme and Mike Karmelich of NYKLauritzenCool USA got me onto the shipping docks and into a Japan-bound freighter.

Tim Gallagher, president and publisher of the *Ventura County Star*, and Joe Howry, the paper's editor, generously allowed me to spend a year working on the series of stories from which this book is adapted. They placed a great deal of trust in my judgment, and backed it with the full resources of the paper. I hope I have repaid their faith.

I also am indebted to those who helped me appreciate the labor that makes farming possible. Henry Vega provided the contractor's perspective, and Lorenzo Vega shared reminiscences about his experiences as a *bracero*. Barbara Macri-Ortiz offered insights from her years as a United Farm Workers attorney and farmworker advocate. Jaime Ceja, the manager at Villa Cesar Chavez apartments in Oxnard, helped locate a farmworker family willing to open their home to me and my photography crew, and provided invaluable translation help. Santiago and Guadalupe Flores allowed us to enter their home, ask them personal questions, and follow them to work so we could get an idea of what it's like to raise a family on a field worker's salary.

Last but not least, I thank Larry Yee, county director of the University of California Cooperative Extension, who for many years has been a quiet but influential voice for local agriculture. A few years ago at a dinner party, he showed me a book about Washington state apple farmers and suggested we produce something similar about Ventura County farming. I did not anticipate that his idea would eventually consume two years of my life, but it did, and his insight and encouragement were invaluable as the long process unfolded.

—John Krist
Ventura, Summer 2007

Pixie tangerines, Churchill-Brenneis Orchard.

Gary Faye

Introduction

It is one of those postcard-perfect winter days that have lured settlers to Southern California for generations, and Jim Churchill is walking through the Ojai Valley orchard he inherited from his dad. He's surrounded by trees loaded with fruit, globular bursts of color scattered liberally amid the glossy green leaves like Christmas ornaments hung with more enthusiasm than discretion.

Most are tangerines of various varieties, along with grapefruits, limes and other types of citrus. As he walks and talks about the vagaries of fruit production, Churchill stops to pluck a bright yellow globe from a drooping branch, and slices it into bite-sized pieces with a pocketknife. "You have to taste this," he says, handing it to a visitor, and then the scene is repeated, each sample unleashing a burst of flavor and a cascade of sticky nectar.

This is the romantic magic of farming, and of citrus farming in particular: sunshine, water and soil transmuted by some sort of biochemical alchemy into mouthfuls of sweet juice, neatly packaged and hanging right there for the plucking. If you could take that blue sky and the winter warmth spilling down over the Ojai Valley on this sunlit February day, squeeze out their essence and distill it, you'd end up with something like a ripe orange or tangerine.

But anyone who's owned or worked on a farm or ranch knows there's very little that's romantic or magical about the day-to-day routine. The work is typically hard, usually tedious and sometimes dangerous. And at the end of the day, the owners of farms and ranches are running manufacturing businesses intended to make money.

Unlike most manufacturers, however, farmers don't control conditions in their factories; those are set by nature, which whimsically doles out profit-killing drought, frost, floods and pests. Farming is like having a crazy business partner who periodically blows the company payroll on lottery tickets.

It wasn't that long ago that most Americans lived on farms and knew these things intimately. Today, however, most people are not farmers, have never worked

Freshly washed lemons, Saticoy Lemon Association packinghouse. *Gary Faye*

on a farm, and have not learned these lessons. Much of what they know about modern farming they've learned by gazing up at green hillsides from their suburban back yards, or by looking through the windshields of their cars as they drive past the furrows at 65 mph.

A STATISTICAL SNAPSHOT

Alone among the coastal counties of Southern California, the landscape of Ventura County remains dominated by farms, a verdant reminder of what the entire region looked like before the sprawling megalopolis washed over the orchards and vegetable fields of Los Angeles and Orange counties like a concrete tide.

Carve off the parts of the county that are sequestered in national forests and parks, and the remainder works out to about 3 acres of farm for every acre of

city: 332,371 acres of agricultural land and just under 100,000 acres of urbanized land, according to the California Department of Conservation's Farmland Mapping and Monitoring Program.

Of the total area in farms, about 100,000 acres is irrigated cropland. So, one way of thinking about the county's landscape is to imagine that for every block of strip mall, city street and housing tract, there's a corresponding plot of celery, strawberries, lemons, peppers, flowers.

Farming is not just a category of land use. According to a recent economic analysis, agriculture supports as many as 47,000 jobs in Ventura County, more than any other sector of the economy except services. It is responsible for nearly 5 percent of the region's total economic activity, spinning off nearly $2 billion a year. Today, Ventura County farmers ship lemons to Japan and truck fresh strawberries to Manhattan. They grow nursery stock, oranges and avocados, peppers, field greens and celery—more than 100 crops in all, from artichokes to orchids, arugula to zucchini.

In general, farms in Ventura County are smaller than farms elsewhere, averaging 143 acres. The statewide and nationwide averages are 346 acres and 441 acres, respectively. But even that figure gives a misleading idea of the size of most Ventura County farms. The median—meaning half are larger and half are smaller—is 20 acres, according to census data. (The Census Bureau defines a farm as any operation that produces at least $1,000 worth of commodities in a year.)

Ventura County farmers tend to be a little older than those elsewhere in the state and nation, and their farms are both more expensive to operate and more profitable. Like many other statistics, however, data regarding average net income hide within them a harsher truth: Although the average profit per Ventura County farm in 2002 was a seemingly comfortable $103,446, more than half of all local operations actually lost money that year, according to federal statistics.

The number of farms in the county has been dropping, from a 1998 peak of 2,760 to 2,318 as of 2002, according to the latest Census of Agriculture. And in

Picking strawberries for processing.

Gary Faye

a recent report for Ventura County's Workforce Investment Board, two researchers—Charles Maxey, dean of the School of Business at California Lutheran University in Thousand Oaks, and Bill Watkins, executive director of the Economic Forecast Project at the University of California, Santa Barbara—warned that rising land and labor costs, and increased competition from foreign producers, likely will cause the number to continue dropping.

"The ultimate result," they concluded, "will be a slowly shrinking agricultural sector. Output will decline, as will the agricultural workforce. Eventually, Ventura County's agricultural sector will be a remnant boutique business, supplying local top-end restaurants, farmers markets, eco-tourism and the like."

Numbers, however, can take you only so far in understanding the agricultural industry that has quite literally shaped Ventura County. To go deeper, you have to spend time with the people who work in that industry.

It's time to meet some of them.

THE BUSINESS OF FAMILY

Joseph Germond Leavens brought his wife, Mary, from New England to Santa Paula in 1900, arriving by train to find it a crude settlement of dirt streets, rough buildings and rougher-looking people. Mary was all for climbing right back on the train and returning to civilization, according to family lore, but was persuaded to stick it out. J.G. opened a dry goods store, and by 1910 had persuaded his brother to invest with him in farmland east of town.

Nearly a century later, the great-grandchildren of J.G. and Mary still farm some of that land, a 43-acre parcel they call Hardscrabble Ranch. It's one piece in a jigsaw puzzle of properties painstakingly cobbled together over the generations, some of it held by a family trust and some owned by the partnership that manages all the properties.

That partnership is now an entity called Leavens Ranches. It's run by Link Leavens, his sister Leslie Leavens-Crowe, and their cousin, David Schwabauer.

The family grows lemons and avocados on about 950 acres in Ventura County, divided among six ranches (they also farm 350 acres in Monterey County).

   Three of the ranches are in or near Santa Paula, one is in Saticoy, one is in East Ventura, and the other is near Moorpark and accounts for more than half the farming operation. David runs that ranch; Link is general manager of Leavens Ranches and oversees the farming operations in the west county; Leslie runs the office and represents the family interests as a member of the board of directors of the Saticoy Lemon Association, the Farm Bureau of Ventura County and other organizations.

   Leavens Ranches headquarters is in a graceful 19th century farmhouse surrounded by 30 acres of lemon trees and 18 acres of avocados off Telegraph Road west of Santa Paula. It was the house where the grandparents of Link,

Leslie and David lived, and there are still a couple of rooms upstairs maintained as bedrooms for visiting family members, many of whom are scattered across the country.

Therein lies a clue to one of the biggest challenges—and highest priorities—for the descendents of J.G. and Mary Leavens: negotiating the treacherous currents at the confluence of blood and business.

"We are committed to being a family, and to farming," Leslie says during a tour of the farmhouse office, made cozy on a drizzly winter day by a fire in the old living room fireplace. "That doesn't just happen. It takes work."

RUNNING WITH THE BULLS

Richard Atmore is a big guy with a big smile and a big handshake. Also, a big pickup truck and a big cowboy hat. He runs R.A. Atmore Ranch which runs a herd of several hundred cows, calves and bulls on about 6,800 acres in the hillsides north of Ventura.

Because the terrain is surprisingly vertical, Atmore and his cowboys still use horses rather than all-terrain cycles or other motorized vehicles to ride the range. This makes the operation something of a throwback in an era of computers and GPS navigation. It's also a direct link to the oldest form of commercial agriculture in Ventura County: the hides-and-tallow business of the Spanish mission and rancho eras.

Rich's ranch is known as a cow-calf operation—he breeds his own cows to his own bulls, and raises the resulting calves until they weigh about 500 pounds. At that point, they're sold at auction to "stockers"—ranchers who specialize in adding another 400 or so pounds to each grass-fed steer—who in turn sell them to feedlots, where the animals typically will be fed grain until they reach market weight of around 1,200 pounds and are slaughtered.

Richard is an engaging man who seems to be having more fun working than most adults do when they're relaxing. He was born in North Carolina but moved

Richard Atmore and friend.

Gary Faye

to Ventura when he was in fourth grade. He learned the ranching ropes, so to speak, from a couple of real old timers, Rocky Esparza and Toots Jauregui, who ranched adjoining leased parcels in the Ventura hillsides. Rich started working on Rocky's ranch in 1979 when he was two years out of Buena High School and bought out the operation when Rocky retired about eight years later. When Toots retired—it was a long wait; he was still riding horses when he was 90—Rich took over the lease to the adjacent property as well.

Ranching on the edge of the city brings challenges no rancho owner ever had

Roundup gear, Atmore Ranch.   *John Krist*

Cecil Martinez.   *Gary Faye*

to contend with. Trespassing hikers cut fences so they can make their way to Two Trees, a local landmark that commands a jaw-dropping view of the Pacific, or to follow trails leading out of Arroyo Verde Park. Rich occasionally gets calls warning him that his cows are running down Foothill Road or that they've broken out to graze the lush grass in the popular park, which holds a potent attraction for his animals in the dry season.

"Feed and girls are two things bulls are going through a fence for," Rich says.

He says lots of things like that. You could scour ranches in Wyoming, Colorado, Arizona, Montana and the Dakotas and not meet a more quotable cowboy.

'A SENSUOUS FRUIT'

Cecil Martinez grew up in the Imperial Valley, the son of farm workers who saved enough money to buy their own land and start their own vegetable farm— no mean feat for laborers making 45 or 50 cents an hour. His dad was born in the Philippines, his mom in California.

Cecil came to the Oxnard Plain in 1968 and found the year-round coastal coolness a welcome contrast to the blast-furnace intensity of desert heat. He started working for a local vegetable and tomato grower in 1968, and shifted into strawberries in 1972, just as they were becoming established as a major commercial crop in Ventura County.

His career as a berry grower encompasses nearly the entire history of the fruit—once a niche product and now the most valuable agricultural commodity in the county—on the Oxnard Plain. He's also played a role in the berry industry's possible future, participating in field trials of alternatives to the controversial soil fumigant methyl bromide. The chemical will soon be banned because it damages the ozone layer, alarming growers who've become reliant on its ability to boost crop yields.

For nearly a decade, Cecil ran his own strawberry farm. But the increasing complexity of the highly competitive business, coupled with a desire to scale back as retirement beckoned, nudged him four years ago into a less stressful job as the district manager for Sunrise Growers.

Placentia-based Sunrise is one of the largest marketers and shippers of strawberries in North America, with operations in Orange County, Oxnard, Santa Maria and the Watsonville-Salinas area. Cecil oversees about 20 independent farmers growing strawberries under contract with Sunrise on 525 acres of the Oxnard Plain.

Cecil is clearly enamored of the berry he's spent half his life growing. He has a big metal strawberry affixed to the trailer hitch of his Honda CRV. He calls berries "a sensuous fruit" and talks animatedly about how easy they are to eat—no peeling, no laborious preparation, just rinse them off and pop them in your mouth.

His enthusiastic testimonials about the ripening strawberries in his fields are interrupted frequently by the chirping BlackBerry in his pocket. Growing strawberries is a high-stakes gamble, with potentially high returns and potentially devastating losses. Berry growers tend to worry a lot, and when they're worried they spend a lot of time on the phone.

CULTIVATING CONNECTIONS

Jim Churchill is probably the only farmer in Ventura County whose resume includes childhood appearances in educational films watched by countless

Jim Churchill and Lisa Brenneis.   *Gary Faye*

American schoolchildren, such classic works as "Wonders in Your Own Backyard" and "Wonders in the Desert." His co-star was his sister, Joan, who grew up to become an award-winning cinematographer and documentary filmmaker.

Their dad, Robert Churchill, was a Harvard Law School grad who turned his back on the legal profession and family expectations to study photography. Eventually, he launched a business making classroom films.

He also bought 40 acres of rocky ground at the east end of the Ojai Valley, where the land begins rising like a gentle wave to lap at the feet of the Topatopa Mountains. He planted avocados, sold off part of the land to help finance his retirement, and

ended up with a 17-acre remnant. That's where Jim and his wife, Lisa Brenneis, now grow avocados, pixie tangerines and other specialty citrus varieties.

Jim is an unlikely farmer. He has an undergraduate degree in philosophy from Cal Poly Pomona, which he attended from 1965 to 1968, and a master's degree from the University of Chicago in social sciences.

"I wasn't raised to do this," he says. "I was raised to do something else."

Lisa also came to farming by a circuitous route. Among other things, she's a film production manager and editor, an alumnus of Churchill Films, and author of the definitive guide to digital editing using Apple's Final Cut Pro software.

When asked to be part of a book about farming, the first thing Jim said was, "I'm really not a very good farmer. What I'm good at is marketing."

It's an instructive distinction, although not entirely warranted; fruit trees do not take care of themselves. But instead of relying on a cooperative such as Sunkist, or another marketing middleman—the strategy pursued by most local citrus growers, including the Leavens family—Jim sells fruit directly to consumers at three farmers markets each week, and also sells directly to wholesalers and retail markets.

He believes in forging stronger connections between those who produce food and those who eat it, and Churchill devotes a lot of time to such things as the farm-to-school programs in Ojai and Ventura, which place the products of local farms in school salad bars and organize "meet the farmer" events intended to teach kids about the sources of their food.

Pixie tangerines are Churchill's signature crop, and the petite orange orbs have their own colorful story. For that, however, you will have to read on.

CROPS GROWN IN VENTURA COUNTY, 2007

Alfalfa

Apples

Arugula

Artichokes

Avocados

Barley

Beans (Lima)

Beans (Green snap)

Bees

Beneficial insects, nematodes and snails

Blackberries

Blueberries

Bok choy

Broccoli

Brussels sprouts

Cabbages

Cantaloupes

Carrots

Cattle

Cauliflower

Celery

Cherimoyas

Cherries

Chicory

Chinese cabbage

Chinese greens

Chives

Christmas trees

Cilantro

Collard greens

Container plants

Corn

Cucumbers

Dill

Ducks

Eggs

Eggplants

Endive

Fennel

Flowers

Gai Lon

Grapes

Grapefruits

Herbs

Hogs

Kale

Kiwis

Kohlrabi

Leeks

Lemons

Lettuce

Limes

Mint

Mizuna

Mushrooms

Mustard

Nectarines

Nursery stock

Olives

Onions

Oranges

Parsley

Peaches

Pears

Peas

Peppers

Pimentos

Plums

Potatoes

Pumpkins

Radishes

Rapini

Raspberries

Sage

Sheep

Spinach

Squash

Strawberries

Sunflowers

Swiss chard

Tangerines

Tomatoes

Turf/sod

Turnips

Watercress

Watermelons

Zucchini

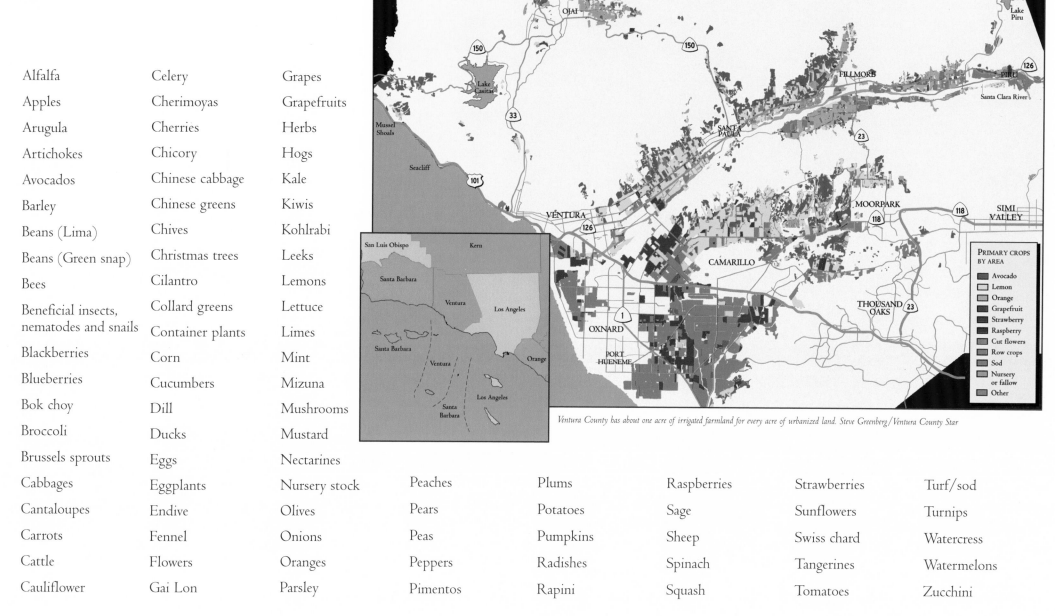

PRIMARY CROPS BY AREA

- Avocado
- Lemon
- Orange
- Grapefruit
- Strawberry
- Raspberry
- Cut flowers
- Row crops
- Sod
- Nursery or fallow
- Other

Ventura County has about one acre of irrigated farmland for every acre of urbanized land. Steve Greenberg/Ventura County Star

15

Winter

Storm clouds over citrus, north of Santa Paula.

Preceding page: New strawberry field, Oxnard Plain.

Gary Faye

CHAPTER ONE
Preparing the Ground

Imagine that it is the spring of 1861, and you are traveling across the Oxnard Plain, making your way from Los Angeles to San Francisco.

Forget about freeways, traffic jams, strip malls and suburban housing tracts. California has been a state for only 11 years and has a non-Indian population of 380,000 people. Los Angeles is a crude, violent pueblo inhabited by fewer than 4,500. San Francisco, more than 12 times as populous, is the queen of the Pacific Coast, made rich and riotous by the river of wealth flowing from the Mother Lode to the Golden Gate.

Here in Ventura County, however, there are more cattle than people and no towns worthy of that title. And this is what it looks like:

> "Friday, March 1, we came on to San Buenaventura, on the seacoast.
> Soon after leaving Cayeguas (Rancho Calleguas, today's Camarillo)
> we entered the plain, which there lies along the sea, and crossed it
> to the sea about twenty miles. It is a fine grassy plain, with here
> and there a gentle green knoll, with a few dry creeks or alkaline
> ponds, and one fine stream, the Santa Clara River, running through
> it. We stopped for an hour on its banks and rested our mules,
> lunched and refreshed ourselves in a grove of cottonwoods which
> came nearer to a forest than anything I have yet seen here. We
> forded the river and came on. At San Buenaventura the hills come
> up to the sea, the plain ceases, but a fine stream comes down from
> a pretty valley, green, grassy, and rich.

> "Here is the old Mission San Buenaventura, once rich, now poor.
> A little dirty village of a few inhabitants, mostly Indian, but with
> some Spanish-Mexican and American. The houses are of adobe,
> the roofs of red tiles, and all dirty enough. A fine old church
> stands, the extensive garden now in ruins, but with a few palm
> trees and many figs and olives—the old padres' garden."

That's how William Brewer described the scene. A child of the East, he had grown up on a farm in upstate New York, later studying agricultural chemistry at Yale and botany in Europe. California's first state geologist, Josiah Whitney, hired him to lead a comprehensive geological survey of California. It was this errand that brought Brewer to Ventura County in 1861.

It is not surprising that Brewer's eye focused on vegetation, wild and cultivated, when he passed through the region. The farm boy who had grown up to become a scientist was convinced that agriculture's future depended on application of the latest scientific knowledge and techniques. It was a conviction he would later put into practice as chair of agriculture in the Sheffield Scientific School at Yale, where he promoted the establishment of agricultural experiment stations and trained successive generations of agricultural scientists and engineers.

Nearly a century and a half later, Brewer's conviction that science and innovation could play a critical role in farming resonates powerfully across the fields, orchards, packing houses and biotechnology labs of Ventura County. The long history of agriculture in this region reflects a continuous pattern of evolution as growers have confronted successive challenges to their profitability—from isolation and aridity to global competition and sprawling urbanization—by adopting the products of research and experimentation.

Yet despite 200 years of continuous change, agriculture in Ventura County still displays some features of its earliest days. To name two: imported water and imported labor. And a third: vulnerability to foreign competition.

These are strange echoes across time, a reminder that 21st century agriculture, despite its embrace of computers, global-positioning satellites and genetic sequencing, sends its roots deep into the past.

IMPORTED WATER

The Chumash, whose ancestors settled in the region at least 10,000 years before the Spanish arrived, did not tend crops but relied instead on wild grass seeds, acorns, tubers, deer, rabbits, fish and shellfish. The food supply provided by nature was sufficiently bountiful to support one of the highest population densities in prehistoric North America, as many as 30,000 people living in 75 to 100 villages strung along the coast between Malibu Canyon and Morro Bay.

True farming began at San Buenaventura Mission, which was established in 1782 near a large Chumash village. The padres and soldiers moving north from Mexico to establish churches and military outposts in California brought with them cattle, grain and other food crops need to make the settlements self-sufficient.

The mission plantings at San Buenaventura were more ambitious than one might conclude from Brewer's 1861 reference to "the old padres' garden." Here's how British navigator George Vancouver described the plantings in their prime, as seen in 1793 during his four-year mapping expedition along the Pacific Coast:

"The garden of Buena Ventura far exceeded any thing of that description I had before met with in these regions, both in respect of the quality, quantity, and variety of its excellent productions, not only indigenous to the country, but appertaining to the temperate as well as torrid zone; not one species having yet been sown, or planted, that had not flourished, and yielded its fruit in abundance, and of excellent quality. These have principally consisted of apples, pears, plums, figs, oranges, grapes, peaches, and pomegranates, together with the plantain, banana, cocoa nut, sugar cane, indigo, and a great variety of the necessary and useful kitchen herbs, plants and roots. All these were flourishing in the greatest health and perfection, though separated from the sea-side only by two or three fields of corn, that were cultivated within a few yards of the surf."

The crops that so impressed Vancouver required irrigation, and the nearest water source was inadequate to the task. To address this lack, the mission work force built an aqueduct of rocks and adobe bricks that stretched for seven miles up the Ventura River valley. The system included a filtration plant and several reservoirs, and provided water to irrigate crops as well as for domestic use in the mission settlement.

So from the first moment that agriculture began in Ventura County, it required imported water and a complex system for moving and storing it—a pattern that would come to characterize the entire state in subsequent decades.

Water was not the only thing the mission fathers imported for their agricultural operations. They also brought with them farm laborers from Mexico, the first example of another pattern that would persist—and stimulate persistent controversy—for the next two centuries.

The mission system gradually disintegrated as the native population of California succumbed to imported diseases and other maladies, and as mission lands were privatized by government decree. By the time Brewer reached San Buenaventura in 1861, only remnants of "the old padres' garden" remained.

Walnut huller, Upper Ojai.

Gary Faye

Earliest archaeological evidence of settled Indian villages in Ventura County. Until the arrival of Europeans, the Chumash and their ancestors gather plants, collect shellfish, fish and hunt animals, but do not cultivate crops.

Spanish government awards first land grant in Ventura County, Rancho Simi, to private owner. Ranchos concentrate on production of cattle for hides, meat and tallow.

Treaty of Guadalupe Hidalgo is signed; Mexico cedes California territory to United States.

Congress passes Land Act requiring California rancho owners to prove title to their land; the process drives many owners into bankruptcy, and most rancho land passes into American hands.

Severe drought kills two-thirds of livestock in Ventura County, forcing many rancho owners to sell.

Lima beans are introduced.

First large-scale orange orchard is planted near Santa Paula.

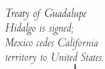

10,000 Years Ago   1782   1795   1834   1848   1850   1851   1861   1863-1864   1867   1868   1871   1874   1876-1877

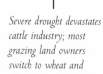

Mission San Buenaventura founded. Using Chumash labor, Franciscan priests establish cultivation of corn, grains, beans, squash, onions, olives, wine grapes and fruit, and raise sheep, cattle and pigs.

Missions are secularized by Mexican government, which distributes church holdings to private owners and encourages settlement in California. Nearly all the arable land in Ventura County eventually is divided among 19 ranchos.

California becomes a state.

Ygnacio del Valle and sons begin experimenting with peach, apple, pear, fig and lemon orchards at Camulos Ranch east of Piru.

Lemons and walnuts are introduced into county.

Avocados are introduced. Hueneme wharf is completed, giving grain growers convenient shipping outlet.

Severe drought devastates cattle industry; most grazing land owners switch to wheat and other grains.

22

Timeline

Southern Pacific Railroad reaches county, sparking land boom and enabling growers to ship perishable crops to eastern markets.

Nathan Blanchard and Wallace Hardison establish Limoneira Co. to raise and market citrus crops on land near Santa Paula; by 1924 it is the largest lemon grower in California.

Gasoline-powered tractors are introduced.

Deepwater port completed at Hueneme.

First commercial strawberry fields are planted.

Ventura voters approve Save Our Agricultural Resources (SOAR) initiative, requiring public vote before farmland in the city can be developed. Over next six years, seven more cities and the unincorporated area adopt similar laws.

Strawberries are ranked No. 1 crop, dethroning lemons from top spot for first time in 54 years.

| 1887 | 1890 | 1893 | 1898 | 1910 | 1914 | 1940 | 1947 | 1957 | 1989 | 1995 | 1999 | 2001 | 2003 |

Grain production peaks in Ventura County, where more than 66,000 acres are devoted to barley, oats, wheat and corn.

Henry, Robert, Benjamin and James Oxnard open the $2 million American Beet Sugar factory at Rancho La Colonia, after local growers Albert Maulhardt and Johannes Borchard demonstrate that sugar beets can be successfully cultivated on Ventura County's coastal plain.

Ventura County Farm Bureau is established.

Lemons become county's top crop.

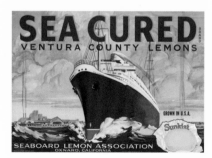

Board of Supervisors establishes Agricultural Land Trust Advisory Committee to study ways of protecting farmland.

Annual value of Ventura County agricultural production tops $1 billion for first time.

Harvested acreage falls below 100,000 acres for first time in more than a century.

Wet field after winter storm, Oxnard Plain.

Gary Faye

It would not be long before ambitious American farmers followed in Brewer's footsteps. In the blink of an eye, they would found an agricultural empire on the "fine grassy plain" watered by the Santa Clara River, an empire that eventually would lay claim to every fertile valley in Ventura County.

A SEA OF GRAIN

If you had stood atop the Conjeo Grade less than 20 years after Brewer's journey and looked either east or west, you would have seen pretty much the same thing:

Grain. Vast fields of it, mostly barley, wheat and corn. Picture Kansas, but with more hills and better weather.

In the years after the Civil War and the California Gold Rush, settlers flocked to Ventura County, where they claimed land as homesteaders, or purchased pieces of the vast ranchos that were being broken up and sold. Severe drought in the 1860s had killed perhaps two-thirds of the cattle and put an end to the rancho economy. Many families were pushed into bankruptcy; others found themselves unable to prove title to lands acquired before California statehood, or incurred crippling debts during their legal battles to establish ownership.

The result was a rapid conversion of Ventura County's "grassy plains" to cultivated cropland. In 1880, according to the U.S. Census Bureau, Ventura County contained 573 farms encompassing 81,107 tilled acres.

By 1890, the amount of tilled land had jumped to 137,349 acres, and more than 54,000 of those acres were planted in barley. Corn came next, at just over 8,100 acres, with wheat planted on about 4,000 acres. The number of farms had risen to 764 and the county's population had nearly doubled, growing from 5,073 to 10,071.

Again, no imagination is required to picture the farms of that era. In 1878 and 1879, the area's two newspapers—the *Ventura Signal* and the *Ventura Free Press*—ran a series of stories that together profiled every farm and ranch in Ventura County. The stories were collected and republished in 2002 by the Museum of Ventura County.

Fruit drying at Robinson Ranch, Upper Ojai, 1890.   *Museum of Ventura County*

Japanese workers at Limoneira Ranch, Santa Paula, undated.   *Museum of Ventura County*

Piru Packinghouse interior, c. 1909.

Museum of Ventura County

Filling sugar sacks at American Beet Sugar Co., Oxnard, undated.

Museum of Ventura County

Here's a sample, a description of Peter Donlon's place near Springville—a long-vanished town near Camarillo:

> "Four hundred acres, all in cultivation, and a good five-board fence with plenty of cross fences. Had in 300 acres of barley, good yield, 15 acres corn. Cut 20 acres of barley for hay. A good alfalfa pasture of 20 acres, has 300 hogs, and good horses and cattle to run the ranch … A fine artesian well supplies water."

It would not be long before Ventura County's grain fields went the way of the vast cattle ranchos that had preceded them. In large part, this would be the result of another factor that continues to influence Ventura County agriculture in the 21st century: global competition.

From barley to berries

Grain prices collapsed in the late 1880s, the consequence of skyrocketing production in the United States and increasing production in Canada and Russia, which displaced American grain from the European market. In response to falling prices, grain production in California plummeted. Ventura County growers seeking better profits replaced grain with beans, a crop that did not require irrigation and could be grown on the same land using similar techniques. Lima bean cultivation in Ventura County rose from 24,000 acres in 1900 to more than 118,000 acres in 1920, its peak year.

Local growers also began raising sugar beets, similarly well-suited to local growing conditions, and as bean prices fell in response to increasing production, sugar beets became a major crop.

This was the start of a pattern that, more than anything, would typify Ventura county farm production in the 20th and 21st centuries: the embrace and then abandonment of one crop after another, in response to changing markets, land prices, mechanization and other factors.

The beet processing plant built in 1889 on the old Colonia Rancho by Henry Oxnard—who also operated two plants in Nebraska and one in Chino—eventually became the second-largest in the nation, under ownership of the American Beet Sugar Co. By the time the plant closed in 1959, it had produced 39 million sacks of sugar, each weighing 100 pounds. The town that grew up around the plant took the name of the plant's founder.

Elsewhere in Ventura County, growers had been experimenting with vegetables and orchard crops, and in the 1920s and 1930s these crops moved to the fore. Saticoy soon dominated California walnut production. Santa Paula-based Limoneira became California's top lemon producer. Lemons became the county's top crop in the 1930s and held that position until 1999, when they were displaced by strawberries, which continue to hold the No. 1 spot today.

The dominance of strawberries—a delicate, highly perishable fruit, which requires enormous investment in land preparation—probably would astonish the farmers of earlier eras, accustomed to cheaper, hardier products such as wheat, beans and half-wild range cattle.

Or perhaps not. One of the earliest written accounts of this region comes from Pedro Fages, a Spanish soldier who was with the first exploratory party to reach San Francisco Bay, and who served as governor after leading numerous expeditions up and down California. In 1775, he penned an entry in his journal describing the landscape of the Central Coast, using words that suggest he would not be surprised at all were he to return today and find the entire region as well-tended as a mission garden.

"It is not to be denied that this land exceeds all the preceding territory in fertility and abundance of things necessary for sustenance," he wrote. "All the seeds and fruits which these natives use, and which have been previously mentioned, grow here in native profusion."

Bean threshing crew, 1909. *Museum of Ventura County*

Harvesting barley at Broome Ranch, 1897. *Museum of Ventura County*

Apricot orchard, Upper Ojai.

Gary Faye

CHAPTER TWO

Natural Advantages

Ventura County growers begin harvesting strawberries in January, when snowdrifts still blanket much of the rest of the country. Avocados are ready to pick from evergreen orchards in late December, when most fruit trees in North America are leafless. Local packinghouses are flooded by a river of ripe lemons in February, when actual rivers in the nation's heartland remain frozen from bank to bank.

That growers in Ventura County are harvesting crops at a time of year when their counterparts in most of the nation are waiting for winter to relax its grip on dormant trees and frigid soil is a testament to Ventura County's extraordinarily temperate climate. Farm production here defies the seasons, turning the traditional progression—spring planting, summer maturity and autumn harvest—completely on its head. In any month of the year, something is being picked or gathered in Ventura County. For some crops, there is almost no halt to the harvest between January and December.

An extended harvest season and the ability to produce multiple crops each year from the same plot of ground give growers here a crucial edge in the cut-throat global competition for consumer dollars, allowing them to move products to market at a time when producers of the same products in less-favorable climates cannot. This enables them to command top dollar and helps explain why the landscape of Ventura County remains dominated by farms despite soaring land prices, rising labor costs and the inexorable press of expanding cities against the rural-urban boundary.

It takes more than just a salubrious climate, however, to establish and maintain an agricultural empire. You also need at least two other key natural resources:

abundant and affordable water, and fertile soil. As growers quickly discovered more than a century ago, Ventura County is unusual in being blessed with all of these key resources, even though it is located in a semi-arid region that barely avoids classification as a desert.

The county owes much of this good fortune to the geological equivalent of a train wreck.

BURIED TREASURE

If you look at a relief map of Ventura County, you will quickly notice that its topography is characterized by a series of parallel mountains separated by valleys, all running roughly east to west. It is as if the county were a rug between a table and a wall, and a clumsy furniture mover had just shoved that table against that wall, crumpling the rug into a series of folds.

In a sense, that's what has happened. Ventura County is riding atop one of several great plates into which the Earth's crust is divided. That chunk of crust, the Pacific Plate, abuts the neighboring North American plate along a suture line known as the San Andreas fault, which slashes through the Tehachapis north of Ventura County. The collision between those two plates has warped the surface of Ventura County like a crumpled rug.

The complex topography of the county, with all its folds, wrinkles and dissections, also affects its climate and weather. There are scores of distinct microclimates in the county, from the marine-influenced plain near the mouth of the Santa Clara River where the temperature hardly varies from month to

Lake Casitas.

Gary Faye

month and strawberries prosper, to hot interior valleys preferred by oranges and tangerines, and mountain slopes dusted by winter snow and grazed by cattle. These microclimates contribute to the wide range of crops produced in the county, more than 100 in all.

As the mountains rise and the valleys sink, erosion tries to even things out. Debris plucked from the high ground by rainfall and runoff fills the sunken basins with a mélange of sand, clay, gravel and cobbles thousands of feet thick. The lowlands, where rivers have spread a fertile frosting of flood-borne silt on this geological layer-cake, are covered with some of the richest soil in the world. There are dozens of distinct local soil types, according to the Natural Resources Conservation Service, bearing such names as Anacapa sandy loam, Camarillo loam, Ojai stony fine sandy loam. In some places, that soil is more than 10 feet deep.

The valleys, pinched between rapidly rising mountains, also channel the creeks and rivers that drain Ventura County, the main systems being those of the Ventura and Santa Clara rivers. Over millennia, rainfall and river water have percolated into the thick deposits of eroded debris filling those basins, turning the porous subterranean layers into the sandy equivalent of a saturated sponge.

These aquifers may constitute the county's most valuable natural resource. According to data compiled by the California Department of Water Resources, the county's major aquifer systems together have a storage capacity of about 21 million acre-feet, and hold an estimated 16.7 million acre-feet of water.

An acre-foot of water is 325,900 gallons, the average annual consumption of two Southern California households. The storage capacity of Lake Casitas, the biggest surface reservoir in the county, is 254,000 acre-feet. Lake Shasta, the largest above-ground reservoir in California, holds 4.6 million acre-feet. Lake Mead, the largest reservoir in the United States, holds 27.4 million acre-feet.

There are many ways to put a price on water, depending on what it's being used for and who is selling it. But farmers buying water from the State Water Project to irrigate crops in the coastal region around Ventura County pay from $394 to

Intake flume for Farmers Ditch, Santa Clara Canal, 1909.   *Museum of Ventura County*

$548 an acre-foot, according to the Department of Water Resources. So, one way of thinking about the prodigious amount of water stored underground in Ventura County is that it would cost between $6.6 billion and $9.2 billion to import a similar quantity from Northern California.

Taking control

When the first settlers arrived on the Oxnard Plain, they found it easy locate water. So much was stored in the great aquifer system beneath their feet that it literally shot out of the ground when they punched holes into it. In the series of farm profiles published in 1878 and 1879 by the *Ventura Signal* and *Ventura Free Press*, the descriptions time and again refer to the abundance of easily tapped water.

Artesian well, Alvord Ranch on the Colonia, undated.   Museum of Ventura County

Zone Mutual Water Co. irrigation well No. 4, 1923.   Museum of Ventura County

Many wells dug by Oxnard Plain settlers were only 10 or 11 feet deep, the newspapers reported.

"Standing back from the road is Mr. (Peter) Donlon's neatly painted two-story dwelling, with a balcony running around the second story," according to one account. "A fine artesian well supplies water both up and down stairs, and is conducted through the stables and lofts."

Many of the early commercial crops planted in Ventura County required no irrigation. Wheat, barley, corn, beans and walnuts, which for decades accounted for the bulk of local production, could be grown even with the somewhat skimpy rainfall of the county's coastal region, about 13 inches a year on average. Shallow artesian wells were adequate for the household gardens and small fruit orchards of the early homesteads.

But as they experimented with new crops and expanded their commercial plantings, growers soon found they needed more water than could be supplied by such wells. They established mutual water companies and dug deeper wells or installed diversion structures on the Santa Clara River to supply extensive networks of ditches and flumes servings thousands of acres of orchards and vegetables.

In 1925, local growers formed the Santa Clara River Protective Association in an effort to retain control over the waters of the river and its tributaries. They were motivated in part by alarm over the actions of the sprawling metropolis to the south: Los Angeles had recently ignited violent conflict with farmers in the Owens Valley by appropriating the valley's water, and the city was building St. Francis Dam on a tributary of the Santa Clara known as San Francisquito Creek—a dam that would collapse three years later, unleashing a flood that killed more than 400 people and denuded thousands of acres of fertile cropland in Ventura County.

But the growers seeking control over the Santa Clara also were motivated by the actions of neighbors closer to home. Residents of the Ojai and Conejo valleys were hatching plans to dam Sespe Creek, the Santa Clara's principal tributary,

Creek and cropland, Oxnard Plain.

Gary Faye

Groomed topsoil, Oxnard Plain.

Gary Faye

and to divert its water through tunnels and pipelines outside the watershed.

The Protective Association evolved into the Santa Clara Water Conservation District in 1927, securing rights to the flow of the river and its tributaries. Wells, however, continued to supply most of the increasing quantity of water needed to irrigate crops and serve the growing cities of the coastal region.

But the quantity of water stored in the county's capacious aquifers is not infinite. It took a long time to fill those aquifers—some of the water beneath the Oxnard Plain has been there for more than 25,000 years, according to the U.S. Geological Survey—and pumping out more than nature puts back each year causes the water level to drop. This can cause wells to run dry, and pumping costs to increase. In extreme cases, it can cause the land to sink and seawater to contaminate the aquifer.

Which is precisely what happened.

FORCING BACK THE SEA

In 1928, the Santa Clara Water Conservation District began replenishing the groundwater basins beneath the Oxnard Plain, constructing spreading grounds and diverting water into them from the river using sand dikes near Saticoy. Despite those efforts, seawater began turning up in wells near the coast in the 1930s.

Efforts to combat the problem and secure adequate water for the region led in 1950 to creation of the United Water Conservation District, a partnership between the old Santa Clara River Protective Association and the city of Oxnard to address both agricultural and urban needs.

United built Santa Felicia Dam in 1955, to capture and store winter runoff in Lake Piru for release down Piru Creek and into the Santa Clara River during the dry season. It also constructed a pipeline to new spreading grounds at El Rio, to increase the amount of river water it could put back into the aquifer.

Still, seawater continued to render more and more wells unusable throughout the 1960s, '70s and '80s, and in parts of the Pleasant Valley area, the land surface

sank more than two feet as the depleted aquifer collapsed. In 1986, United constructed a pipeline to deliver diverted river water to farmers on the Oxnard Plain and reduce their groundwater pumping. In 1991, the district installed the Freeman Diversion to shunt additional storm flows from the river to recharge groundwater basins. Limits also were placed on private pumping. For the most part, the strategies have succeeded in reversing the seawater intrusion, although some wells remain contaminated.

While farmers and cities were working to conserve groundwater and keep the Santa Clara River in local hands, the federal government was helping develop the Ventura River watershed.

Established in 1953, the Ventura River Municipal Water District approached the U.S. Bureau of Reclamation with a proposal to share the costs of developing a project in the watershed to serve the growing city of Ventura and farmers in the Ojai and Ventura River valleys. Authorized by Congress in 1956 and completed in 1959, Casitas Dam and reservoir are now operated by the Casitas Municipal Water District, successor to the Ventura River Municipal Water district.

Also in 1953, residents of the eastern and southern parts of the county formed the Calleguas Municipal Water District, which joined the Metropolitan Water District of Southern California to import Northern California water. That water is delivered mainly to urban customers, reducing competition between cities and farmers over local sources.

MONEY AND POWER

Just as growers in Ventura County have traded one crop for another over the past century—grain and beans replaced by nut and fruit orchards, then vegetables, and eventually berries and nursery stock—so, too, have they adopted new irrigation methods in response to changing conditions.

The lemon and avocado trees managed by Link Leavens, his sister Leslie Leavens-Crowe, and their cousin David Schwabauer are kept green by mini-

New avocado orchard, Moorpark.

Gary Faye

sprinklers on foot-high risers attached to black plastic tubes snaking through the orchards. So are Jim Churchill's Ojai pixie tangerines.

In the Sunrise Growers strawberry fields overseen by Cecil Martinez, long lines of dripper tape laid down below weed-inhibiting plastic sheeting keep the berries flourishing by delivering regular, carefully calibrated doses of water right to the roots.

In the old days, strawberry fields—like many fruit orchards and other crops—were irrigated by flooding them, Martinez said. That wasted a lot of water and water is expensive even if it's being pumped from a well rather than imported at great cost from hundreds of miles away. Much of the expense is related to the prodigious amount of energy gobbled by the pumps that feed the irrigation systems.

In the Leavens family's Moorpark avocado orchard, for example, the trees are irrigated using groundwater. The well shafts reach as deep as 1,200 feet, and draw water from about 500 feet below ground. Some of that water then has to be pushed 300 feet up a hill to irrigate the highest plantings.

During peak summer usage, said Schwabauer, who manages the family's Moorpark operation, the electricity bill for running the irrigation pumps is $30,000 a month.

"We're growing a tropical fruit in a semi-arid environment," said Leavens, pointing to the avocado's origins in the rain-drenched jungles of Central America. Not irrigating is not an option, he said—not if you want the fruit to gain size and the shallow-rooted trees to remain green when the weather is hot and dry.

Grazing cattle also need water, and in the hillsides north of Ventura there is not much flowing above ground during the dry season. Rancher Richard Atmore pumps water uphill from a well at the edge of Arroyo Verde Park to a huge tank atop a ridge with a million-dollar view of Ventura, the Pacific and the Channel Islands. Pipes run from there to fill stock tanks in the canyons below, without which his cows would have a hard time finding a drink in June and July.

Ground preparation west of Santa Paula.   *Gary Faye*

Even with the availability of abundant groundwater and captured runoff impounded behind dams, many Ventura County growers still look on water falling from the clouds as a welcome gift. Irrigation tends to cause salt to build up in the soil, eventually damaging plants and reducing crop yields. A good 2-to-3-inch rainfall, however, flushes the salt from the soil.

It also saves the growers the cost of pumping water from deep below ground, up hills and across their fields—at least for a few days.

"When it rains," Leavens-Crowe said, "it's like hundred-dollar bills falling from the sky."

Ventura County Farming by the Numbers

Land in farms: 332,371 acres

Percentage of county land in farms: 28.1

Irrigated farmland: 99,866 acres

Prime farmland: 47,877 acres

Prime farmland converted to nonagricultural uses 1992–2002: 3,695 acres

Number of farms: 2,318

Number of registered organic growers: 47

Average farm size: 143 acres

Statewide average: 346 acres

U.S. average: 441 acres

Average age of farm operator: 58

State average: 56.8

U.S. average: 55.3

2005 crop value: $1.2 billion

2005 value of top crop (strawberries): $328.6 million

Ranking for crop value among California counties: 9th

Rank among all U.S. counties: 10th

Average annual sales per farm: $439,544

California average: $323,205

U.S. average: $94,245

Average annual production expenses per farm: $348,740

California average: $257,701

U.S. average: $81,362

Average net profit per farm: $103,446

California average: $74,469

U.S. average: $19,032

Spring

Celery field, Oxnard plain.

Preceding page: Strawberry field at height of the season.

Gary Faye

CHAPTER THREE

Picking and Packing

This is a not a particularly good day to be a calf on the R.A. Atmore Ranch. A couple hours after sunup, five men on horseback ride into a corral tucked within the hills and canyons north of Ventura. Bawling cattle mill about inside the enclosure, which is welded together out of salvaged oilfield pipe and appears sturdy enough to stop or at least deflect a runaway 18-wheeler. About half the captive animals are mother cows and the other half are their calves; all were roaming the green springtime hills in feral tranquility until two days ago, when cowboys began rounding them up and herding them here.

Over the next hour, ranch owner Richard Atmore and the other mounted men separate the cows from the calves, driving the reluctant mothers into one pen and their alarmed offspring into another. Then, a few at a time, the calves are driven back into the main corral, where they are roped, flipped, pinned to the ground, branded and vaccinated. Notches are cut in their ears, budding horns are lopped off, and those calves that began the day as young bulls are castrated.

The process generally takes less than three minutes, although they are noisy, dusty, frantic minutes. Wobbly and bleeding, the calves are then chased into the next pen to rejoin their mothers and perhaps wonder with dazed anxiety what further unpleasantness the afternoon might bring.

Branding, an early milestone in the process that turns sunshine into grass and then into plastic-wrapped slabs of sirloin, is one of the rituals of spring in Ventura County's agricultural landscape. Spring is also harvest time for many of the county's leading crops, a profound rearrangement of the traditional seasonal progression—spring planting, summer maturity and autumn harvest—and

Roped calf undergoes the rigors of branding day.   *Karen Quincy Loberg/Ventura County Star*

testament to the region's beneficent climate. In April and May, pickers swarm local fields and orchards, trucks loaded with cartons of vegetables and bins of fruit crowd the highways, and packinghouses rattle and rumble with activity.

Picking and packing. That's as good a way as any to sum up springtime on Ventura County's farms and ranches. In Jim Churchill's tangerine orchard, in the Leavens Ranches avocado and lemon groves, and in Cecil Martinez's strawberry

Picking strawberries for processing.

Gary Faye

fields, ripe fruit is beginning its journey to market.

A journey is beginning for the calves on Atmore's ranch, too, although poking and prodding better describes their experience.

IN THE PICKERS' HANDS

This is a perfect strawberry.

It is big and plump, about the volume of a pingpong ball, and symmetrically conical, tapering to a distinct point. It is bright red and shiny, not the deeper and duller hue that signifies dead ripeness and short shelf life. It bears no hint of green and it is unblemished: no bruises, scrapes, nicks or nibbles.

The taste of that perfect strawberry? Not so important. When the fruit intended for large commercial buyers is graded, there is neither a taste test nor any evaluation of sugar content and acidity. Sweetness can actually be a liability for well-traveled berries, because high sugar content accelerates the inevitable process of decay. For the mass market, what counts is appearance, and appearance means money. Growers often plant sweeter varieties for sale locally, whether at roadside stands, farmers markets or small retailers that deal directly with farmers.

A top-grade load of berries will bring the grower as much as $2 more per tray than the next-highest grade, Martinez says. A tray holds 8 to 10 pounds of berries, and growers in Ventura County can expect to pull as many as 5,000 trays of strawberries out of each acre of fertile ground over the course of an average harvest season, which begins in January and ends in mid-July. The difference between top-grade and lower-grade fruit could theoretically mean $10,000 an acre—a million bucks a year in gross revenue to a grower farming 100 acres.

The grade given a particular load of berries is in part due to the diligence of the farm owner or manager, who must juggle a host of factors—rain, soil fertility, temperature, assault by an army of pathogens and pests—to maximize both total yield and the percentage of each plant's production that can be sold for top dollar.

"As a grower, you come to look at the plants every day," Martinez says while walking through the rows. Out in the field, which smells sweetly of the ripe red fruit, Martinez can detect the subtle clues in leaf color suggesting a need for more water or fertilizer, determine whether mold or rot is gaining a foothold, or whether pests are invading one of his meticulously groomed fields.

But the grade and therefore profitability of a load of berries depends ultimately on the skill of the harvest crew.

The pickers crouch in knee-deep furrows that separate each bed of plants from the next. Each bed has four parallel rows of plants, each plant bearing fruit at various stages of maturity, from just-fertilized flower to bright red berries. Each picker pushes a small wire cart supporting a cardboard carton filled with clear plastic clamshell containers.

Typically wearing thin gloves, the pickers run their hands through the plants, brushing back the leaves to uncover the fruit. All the ripe berries must be picked, even those that are too small, weirdly shaped or otherwise defective; left on the plant they eventually will decay and spread rot to other berries.

Pickers do not simply grab a berry and yank; that would bruise the fruit or even pop the stem cap out, opening a route for decay and earning the fruit a low grade. The proper technique is to grasp each berry gently but firmly and then flick the wrist to snap the stem, which produces a clearly audible *snick*. The pickers do this with incredible speed, their hands a constant blur of motion, filling a tray every 10 minutes or so and earning about $10 an hour. On a quiet morning, as a crew of 30 or more pickers passes through a field, the fruity air is filled with a continual *snicksnicksnicksnick*.

Pickers toss unwanted fruit into the furrow and place those berries they judge acceptable directly into the plastic containers. What this means is that the pickers are not just evaluating and picking fruit; they also are packing and preparing the berries for display on supermarket shelves.

Truckloads of filled boxes stacked on wooden pallets are hauled directly from

Strawberry harvest crew, Oxnard Plain.

Gary Faye

the fields to a cooler. In the case of Sunrise Growers, that cooler is at the giant Boskovich Farms packing facility in Oxnard. There, the berries are unloaded and samples are taken from each pallet to be graded by a team of women trained and employed by Sunrise, who wear surgical gloves and a look of careful concentration.

Forklifts then move the pallets of berries into a refrigerated building, where they are lined up against a device called a "tunnel," which uses huge fans to suck chilled air through the stacked boxes. In two or three hours, it can cool the berries from field temperature to 34 degrees, prolonging their shelf life.

The berries than are stacked in a storage area and held at the same low temperature. Those destined for distant markets are rolled into another machine, which seals the entire pallet load in a plastic bag and pumps it full of a proprietary gas blend—mostly carbon dioxide—intended to retard mold and spoilage. From storage, the pallets are loaded directly into the long-haul refrigerated truck trailers that will carry them to market.

A tray of berries picked this morning in Oxnard can be in a Portland health-food store tomorrow night. In 72 hours, a Cajun grocer can be stacking those berries on a display table in Louisiana.

Top grade, top dollar

This is a perfect lemon.

It is the size of a small fist, cylindrical in the center, tapering slightly toward each end. It is hard, with no give to it when squeezed, and weighs 5 or 6 ounces. It is of a size, shape and heft that practically invite you to see how far you can throw it. The skin is lightly dimpled but not rough, and it is a uniform pale yellow. It has no scars, no blemishes of any kind.

"A lemon is a lemon is a lemon? Not really," says Link Leavens.

There are four lemon grades. The fancy-grade lemon gets to wear the Sunkist label and brings the highest price. Next is choice grade, generally destined for food service customers—decorative garnish on the buffet table, a wedge in the

restaurant glass of iced tea. Third is standard grade, typically used for lemonade, juice concentrate and flavoring for other beverages, such as Coca-Cola. On the bottom rung are culls, which end up as cattle feed.

More Ventura County acres are planted in lemons—20,875 in 2005, according to the Agricultural Commissioner's Office—than any other crop. Forty to 50 percent of each year's lemon production is sold for juice, concentrate and other products, at a price that doesn't cover the cost of producing and processing it. About half is sold on the fresh market and earns growers a profit. Perhaps 15 percent of the total crop is top grade and brings the highest price.

Growers can influence the quality of the crop by applying the right amount of water and the right blend of chemicals at the right time. Leavens, for example, keeps computerized records of the nutrient content of leaves and stems, which he can compare to his records of fertilizer application to make sure he's getting the biggest bang for his chemical buck. The quality of the crop also depends on the skill of the members of the pruning crew, who trim and remove branches to allow sunlight to penetrate to the fruit-bearing wood, and to force the tree's energy into producing fruit rather than greenery. Much also depends on the whims of nature—fruit scarred from being banged around by wind brings a lower grade, as does fruit disfigured by thrips, mites and other pests.

The difference in price between one grade and another can be $4 to $5 per 40-pound carton. When you produce as many lemons as the Leavens family does—about 3.6 million pounds a year—that adds up. As is the case with strawberries, the pickers' skill can powerfully influence a grower's profit margin.

The pickers in the Leavens Ranches orchard have heavy nylon bags slung around their necks. Each picker carries a set of palm-sized shears with tiny blades. From a distance, the shears are invisible, so neatly and completely are they enclosed by a cupped hand.

Yanking a lemon off the tree can damage the branch or pull the stem bud out of the fruit, which provides entry to bacteria and accelerates the ripening process,

Sorting fruit, Saticoy Lemon Association packinghouse.

Gary Faye

so picking is a two-handed operation: The picker uses one hand to grab the piece of fruit and deftly swipes at it with the hand holding the clippers, snipping the stem just a fraction of an inch from the lemon's tip. The hand holding the fruit then flicks it into the sack.

This is done so quickly, and with so little wasted effort and motion, that from a distance it appears that the pickers merely wave their gloved hands across the thorny branches, which magically send a continuous stream of fruit arcing into the bag.

When the bag is full, the picker walks to the nearest bin and carefully dumps the 50 or 60 pounds of fruit into it. A good lemon picker can fill five or six bins a day. Each bin holds nearly a half-ton of fruit and earns the picker $20 to $25, depending on the season.

Most Leavens lemons are trucked to a Sunkist-owned Saticoy Lemon Association packinghouse. (Some overflow is handled by the Limoneira house in Santa Paula.) Each year, the association processes about 480 million pounds of Ventura County lemons, and ships about 260 million pounds to buyers.

A packinghouse smells like bleach and lemon oil. Arriving bins of fruit are loaded by forklift into a machine that dumps them into the wash line, where the lemons float through a warm bath containing cleanser, chlorine to kill mold spores—hence the bleach smell—and a water-soluble wax that will coat the fruit and prevent it from drying out in storage.

Parallel conveyors then shuttle the lemons past the unblinking eyes of computerized optical scanners, which quickly (nine lemons per second per line) assign a grade to each piece of fruit and instruct the conveyor to kick it into the appropriate plastic box.

The boxes then shuttle off to storage. Lemons are picked year-round, and the peak harvest is in spring. Peak demand, however, does not come until summer, when people are barbecuing and drinking lemonade. So, lemons spend an average of 45 days in storage before they're pulled out and fed into the pack line.

On the pack line, the lemons are again washed and waxed, sorted by color and size, and packed into boxes or bags, which then are moved to cold storage to await shipment.

According to Glenn Miller, president of Saticoy Lemon Association and manager of its packinghouses, recent changes in retailing—particularly the growing importance of discount warehouse stores—have meant new kinds of packaging. The Saticoy packinghouse, for example, puts lemons in 28 different cartons and five kinds of bags, which change in response to various marketing campaigns.

"They've got Nemo bags, they've got Elmo bags," Miller says, without speculating as to how a Muppet might boost lemon sales or whether young fans might be appalled by an implied product tie-in between citrus slices and Nemo fillets.

FRESH IS BEST

This is a perfect pixie tangerine.

It is dark orange, and about the size of a golf ball. Or, it's pale orange and about the size of a tennis ball. It's smooth and symmetrically spherical. Or it's knobby and a bit lopsided.

Grading tangerines.   *Karen Quincy Loberg/Ventura County Star*

Pre-dawn sorting at Villa Park Orchards Association packinghouse, Fillmore.   *Karen Quincy Loberg/Ventura County Star*

Unlike oranges, lemons and most other citrus varieties, pixies defy standardization. No two are alike. As packed by Ojai Valley growers such as Jim Churchill, there are only two grades: pass and fail.

Picking tangerines is just like picking lemons—the same techniques, the same tools. Churchill sends tangerines from his 17-acre citrus and avocado orchard to two packinghouses. One is small, operated by fellow pixie grower Mike Shore, and handles loads for relatively small customers. Larger shipments are processed on a special line in a corner of the cavernous Villa Park Orchards Association packinghouse in Fillmore.

On this day, tangerines are being graded and packed in Fillmore for customers in Detroit, Canada, Los Angeles, San Francisco and Sacramento. Later in the week, pixies will be packed for a buyer in Japan.

The pixie line is a simplified version of the lemon line: A forklift delivers each

Pixie tangerines, Churchill-Brenneis Orchard, Ojai.   *Gary Faye*

bin to a hydraulic lift, which dumps the contents onto a conveyor. The fruit rolls past four gloved women, the graders, who pick through the tangerines and cull those that are of poor color, bruised, heavily scarred or damaged by pests. The pixies roll through a set of brushes that remove any clinging debris, pass another set of graders, and then move through a set of optical scanners that sort them by size.

Thus sorted, they roll down a chute and land in front of the packers, who scoop the fruit by hand into cartons. Other workers take the filled boxes to be sealed with tape and then stacked on pallets. The tangerines are not washed or waxed or treated with any chemicals.

One consequence of the pixies' quirky individuality is that they appeal mainly to specialty markets and to customers who know enough to seek them out—a situation perfectly suited to small farmers such as Churchill and Shore, who lack the marketing reach of Sunkist and other large companies or grower cooperatives.

It also means the growers, with limited storage and a customer base motivated almost exclusively by flavor, must move the fruit to market as quickly as they can.

"Part of what we're trying to do is get the fruit to the consumer as fresh as possible," Churchill says. "Fresh fruit really is best."

LIVING HISTORY

This is a perfect steak.

It is deep red, tender and marbled with snow-white fat, which adds flavor and succulence to the meat when it's cooked.

A steer that spends its life roaming the hills and eating only grass does not produce that kind of meat. Range animals are typically lean, and what fat they have is yellowish. A growing number of shoppers seek out grass-fed beef, particularly if it has been raised organically, but to get that tender, white-marbled meat preferred by most consumers, beef producers typically confine steers in feedlots during the last few months of their lives and gorge them on grain.

Eventually, most of the calves born last winter on the R.A. Atmore Ranch will

Roping calves on the Atmore ranch, Ventura.   *John Krist*

find their way to feedlots. For now, however, they're mowing through the luxuriant grass that sprouted on local hills after the late-spring rains.

In January and February of 2006 it did not look as if that would be the case. Thanks to a dry winter, the hillsides were covered with only a weak stubble. Atmore was warily monitoring weather reports, preparing for the possibility that he'd have to sell all 300 or so of his cows rather than try to graze them on inadequate forage, which would strip the hillsides bare and possibly inhibit regeneration in future years.

Cattle ranchers are basically grass farmers. They have to be willing to match herd size to the carrying capacity of the range, Atmore says, or they pay the price in the long run.

"Don't fall in love with your cows," he says. "If you don't take care of the grass, it hurts you."

Richard Atmore makes his mark.

John Krist

By mid-April, repeated storms had fomented a riot of greenery and put a smile on Atmore's face.

"When it gets warm like this after a rain, you can hear the grass growing," he says on one particularly sun-drenched day. With plenty of grass, the cows could stay. So there are plenty to round up and bring into the corrals for branding, vaccinating and castrating.

Gathering cows in the rugged terrain between Ventura and Ojai is harder than it sounds.

"You don't just get behind the cattle and think you're going to end up down in the corral," Atmore says.

This turns out to be a bit of an understatement.

On this morning in late April, Atmore, four other cowboys and one visitor head into the hills on horseback, intending to drive about three dozen cows out of hiding and into the corral near the ranch office in Sexton Canyon. It isn't long before all the riders are crashing cross-country through brush taller than a man in pursuit of recalcitrant cows and calves that behave more like mountain goats than domesticated bovines.

The 6,800-acre ranch is a labyrinth of canyons and gullies eroded into the steep hills. There are a few dirt roads, but the cattle aren't eager to use them.

To keep the cows moving in the right direction means moving swiftly to head off each of many escape attempts, during which the cows often charge straight up slopes that seem nearly vertical, or circle above the cowboys and try to run back up the canyon.

"Richard's cows, they are wild sonsabitches," is how Jorge Casian, foreman of a nearby ranch, puts it during a lull between stampedes.

A few renegade beasts escape to graze another day, but the cowboys eventually manage to dislodge most of the cattle from their hiding places. Altogether, it takes six riders three hours to corral about 20 cows and calves, a rather discouraging ratio of man-hours to cattle.

New planting, Las Posas Valley.

Gary Faye

The next morning is not pleasant for the calves, a bewildering blend of blood, dust, smoke and slobber, foreshadowing the inevitable fate most of Atmore's steers and heifers will face. But it is fiercely evocative of another time and a way of life that has all but vanished: the quick-stepping horses, the whistling lariats, the cowboys in chaps and spurs and well-worn boots cracking jokes in a mixture of Spanish and English, the spectators perched atop fence rails—the scene could well have been lifted from one of the earliest pages of local history.

Except for the Honda generator powering the electric branding iron. Cattle ranching may be a holdover from the oldest form of commercial agriculture in Ventura County, but even it has adapted to changing times. A little bit, anyway.

Young bell pepper plant.

Gary Faye

CHAPTER FOUR
Working the Land

Four-year-old Jessica Flores is racing around with only stockings on her feet, a shoeless whirlwind with long dark hair and a blue sun dress. Chattering happily in Spanish, she darts back and forth between the family's living room and the courtyard outside.

In the small kitchen of the four-bedroom apartment, a saucepan of squash simmers on the stove, lending the air a musky vegetative scent. Having turned over the comfortable sofa and matching chair to visitors, Jessica's parents, Santiago and Guadalupe Flores, occupy two of the four dinette chairs in the small eating area, its floor protected from spills by sheets of plastic, and smile indulgently at their youngest daughter's antics.

"What I like best about this place is that it is calm," Santiago says through an interpreter. "There's a lot of space where the kids can play freely."

That wasn't the case in the home the Flores family had occupied a few months earlier. All nine of them had been crammed into a one-bedroom apartment in Oxnard, and there had been no place outside for a child to play safely unsupervised and unshod. Now, however, they're among 52 families to find lodging in a new low-income apartment complex built by the nonprofit Cabrillo Economic Development Corp. Children are everywhere in the complex, which has the congenial and close-knit feel of a small village.

The Flores family thus straddles two sides of a profound divide.

Like all their neighbors in Villa Cesar Chavez, Santiago and Guadalupe are farmworkers, members of the large low-wage labor force that helps maintain Ventura County's status as one of the leading agricultural regions in the nation. It is a labor force that for the most part endures poverty and substandard living conditions, while making it possible for consumers to enjoy cheap food, local growers to sell a billion dollars' worth of crops each year, and suburban homeowners to enjoy verdant views of orchards and fields surrounding their communities.

That Santiago, Guadalupe and their children now find themselves living in clean, safe and relatively spacious surroundings, rather than the cramped, dilapidated quarters that Ventura County farmworkers more typically occupy, is testament to the power of hope, hard work and dogged efforts by the lawyers, activists and others who seek improved living and working conditions for immigrant laborers.

At the same time, the Flores family's experiences underscore just how rare such good fortune is. And their lives throw into stark relief the painful paradox at the heart of the county's agricultural empire: Ventura County farmworkers make less money than any other category of workers in Ventura County, yet their labor maintains the bucolic landscape that makes the county such a desirable— and extremely expensive—place to live. Raising a family under such circumstances is a daily struggle.

Flores puts it in terms any parent can understand. He looks at his children and says simply, "I don't want them to have the same life I had."

THE FIRST WAVE

As Mexican immigrants, Santiago and Guadalupe are part of a vast influx of job-seekers from Latin America that over the past half-century has reshaped the demographic, cultural and political landscape of the American Southwest.

Celery harvest, Oxnard Plain.   *Gary Faye*

The implications of that influx dominate political discourse today, as activists, lawmakers, commentators and academics debate the nation's immigration policies and border security.

There is nothing new, however, about the controversy over foreign farmworkers. Today's arrivals are links in an unbroken chain of immigrant laborers drawn or driven to California's rural landscape over the past two centuries. From its earliest days, the state's agricultural industry has relied on low-wage workers from distant lands, and for nearly as long, those immigrant workers have been the focus of conflict.

Agriculture arrived in California with the Spanish missions, established between 1769 and 1823 along the coast from San Diego to Sonoma. As the first group of soldiers and Franciscan priests moved north from colonial settlements in Baja California, they brought livestock, grain, fruit and other crops needed to make the new mission settlements self-sufficient. They also brought with them several dozen Christianized Indians from the Baja peninsula to plant, tend and harvest the crops until converts could be recruited for that work among natives in the north.

These imported workers were considered so vital to the survival of the settlements that Junipero Serra, founder of Mission San Buenaventura and head of the mission system in California, traveled to Mexico City to plead with the colonial government to send him more.

Santiago and Guadalupe Flores at work with daughter Maribell, at left.   *Karen Quincy Loberg/Ventura County Star*

"Using arguments that have since become standard for those lobbying for foreign farmworkers, Serra predicted dire consequences if he did not get the extra manpower he requested," Richard Steven Street writes in *Beasts of the Field*, his exhaustive history of immigrant farmworkers in California. "Because of the lack of field labor, he reported, crops had dwindled or not been planted at all, and for this reason he had failed to gather large numbers of natives at the missions."

Serra's plea for new recruits fell on deaf ears, and the number of transplanted farmworkers dwindled over time. California Indians took their place, but within a few decades the mission era ended. California's agricultural economy became one devoted almost exclusively to large ranchos raising half-wild range cattle for hides and tallow, which required little in the way of labor.

Celery harvest crew.

Gary Faye

The state's next significant wave of immigrant farm laborers began arriving shortly after gold was discovered on the American River in 1848. As fortune hunters swarmed into California, business boomed and so did the demand for low-wage labor. Employers began recruiting Chinese immigrants, who played an important role in the state's labor force for the next three decades.

By 1880 there were more than 75,000 Chinese laborers in California. They were critical to construction of the transcontinental railroad as it climbed over the rugged spine of the Sierra Nevada, but they also were instrumental in establishing the wine-grape industry in Napa and Sonoma counties. In Sacramento, San Mateo and Alameda counties, Chinese immigrants accounted for between a quarter and half of the farm labor force.

The backlash was not long in coming. Native workers viewed the immigrants

Lemons head toward the computerized grading line.

Gary Faye

as a threat, believing they drove down wages and took jobs from Americans—themes that would echo across the decades. Starting on the West Coast, anti-immigrant sentiment spread, often taking violent form and culminating with passage in 1882 of the Chinese Exclusion Act, which formally outlawed immigration into the United States from China.

LOOKING TO JAPAN

As labor shortages began interfering with crop harvests up and down the state, California growers turned to a replacement source of workers. By 1909, there were about 30,000 Japanese immigrants working on California farms, accounting for 42 percent of the agricultural labor force, according to the California Bureau of Labor Statistics.

Like the Chinese before them, the Japanese were soon the target of anti-immigrant violence and political reprisals. Some of the hostility was a response to their success at bargaining collectively for higher wages, but some was a direct reaction to their use of farm labor as a steppingstone to farm ownership.

By 1910, Japanese farmers were growing 88 percent of California's strawberries, 60 percent of the state's sugar beets and cantaloupes, 51 percent of the table grapes, and most of the tomatoes, onions and celery. They owned or leased about 13 percent of the state's farmland, and accounted for 21 percent of the total cash value of California farm production, according to historian Street.

In response, the California legislature in 1913 passed the first of several Alien Land Acts, prohibiting noncitizens—which, under federal law, included all Asian immigrants—from owning land or leasing it for more than three years.

Immigrants of a different kind flocked to California's agricultural regions in the 1930s, as the twin calamities of drought and the Depression displaced thousands of tenant farmers and small landowners from the Great Plains. More than 180,000 of these Dust Bowl refugees arrived in California between 1935 and 1939, flooding the agricultural labor market, driving down wages, and temporarily silencing the perennial growers' lament about a looming shortage of workers.

The labor surplus did not last long, however. Following the attack on Pearl Harbor in 1941, which precipitated America's entry into World War II, more than 110,000 people of Japanese ancestry in California, Oregon and Washington were deemed a potential security threat, forced from their homes and jobs, and sent to internment camps. That exodus was accompanied by the sudden loss of much of the remaining work force, as able-bodied men enlisted in the military or abandoned the rigors and paltry pay of field work in favor of well-paying jobs in factories serving the war effort.

In response to the wartime labor shortage, growers persuaded the U.S. government in 1942 to approve the first in a series of international agreements allowing Mexicans to enter the country temporarily to work in the fields.

These *braceros*—the word derives from Spanish "brazo," meaning "arm," and refers to farmhands or hired laborers—signed contracts for terms that ranged from four weeks to six months, promising them the prevailing wage in the area where they'd be working, free housing, reasonably priced meals, employment or subsistence payments for at least 75 percent of their time in the United States, and free transportation from and back to the recruitment center in Mexico. They were supposed to be used only in areas of certified domestic labor shortage, and were not to be used as strikebreakers.

Many of the rules were routinely broken, and such abuses eventually led Congress to discontinue the program. By that time, however, immigration from Mexico—legal and illegal—had begun to soar, a phenomenon some historians have linked to the networks of cross-border relationships and contacts established by the 22-year bracero program.

BACKBONE OF THE INDUSTRY

There is little in the way of reliable data describing the local farmworker population today, an elusive target for demographers, census enumerators

Grading avocados, Calavo packinghouse, Santa Paula.   *Gary Faye*

and economists by virtue of the social, linguistic and legal barriers that keep immigrants in the shadows despite their importance to the county's economy and culture. Even such a basic fact as the number of workers employed in the county's fields and orchards is hard to pin down; the best estimates are just that, estimates, extrapolated from multiple and sometimes conflicting sources.

But by drawing on many of those sources, some of them focusing on statewide demographics, it is possible to paint with reasonable confidence a portrait of Ventura County's farmworkers. These sources include the National Agricultural Workers Survey conducted by the U.S. Department of Labor, data gathered by the California Employment Development Department, and reports by University of California's Giannini Foundation of Agricultural Economics.

There are about 20,000 Ventura County farmworkers, although the number ranges seasonally from a low of 15,000 to a high of 25,000 during the peak spring and summer harvest of strawberries, lemons and avocados. If they are like the rest of California's farmworker population, 95 percent were born outside the United States, 91 percent in Mexico, and at least 57 percent of them are in the country illegally.

The vast majority, 73 percent, are male, and their median age is 32. Half have less than a sixth-grade education, and 53 percent speak no English. Nearly half have been in the United States more than 10 years, and they've been employed in agriculture an average of 11 years.

An increasing number do not speak Spanish or English. Of those California farmworkers who've arrived within the past two years, 38 percent are members of indigenous cultures from southern Mexico. They are survivors of the pre-Columbian population, descendents of the Maya, Aztec and other groups typically found today in remote rural areas. They generally are illiterate in any language, and speak dialects with no linguistic relationship to Spanish or English, such as Mixteco, Zapoteco, Amuzgo and Nahuatl.

LOW PAY, HARD WORK

As for income, it is likely that Ventura County's farmworkers make, on average, a bit more than their counterparts in other parts of California, owing to the longer growing season and multi-crop production that characterizes the coastal region.

Statewide, 75 percent of farmworkers earn less than $15,000 a year and 43 percent make less than $10,000 annually. In part, this is a consequence of low pay: the median farmworker wage in California is less than $8 an hour. But it is also a result of the seasonality of employment: Only 20 percent of farmworkers are employed year-round.

A study prepared in 2002 for the Ventura County Board of Supervisors found that median household income for local farmworkers was $22,000, although that included the earnings of all related people living in the same dwelling. On average, there were five related people living in each unit. The median personal income was reported as $11,758.70.

At the time, the average rent for a two-bedroom apartment in the Oxnard-Port Hueneme area was $1,259 a month, or $15,108 a year. The federally defined poverty level for a household of five that year was $21,744. If the county survey was accurate, at least half of the county's farmworkers lived in officially defined poverty in 2002. (In 2006, the federal poverty threshold was $24,059 for a household of five that included three related children under 18 years of age.)

According to that survey, strawberry workers had a median annual income of $8,000, the lowest of any crop, whereas sod-farm workers made the most, a median income of $33,200. Again, this may well be a reflection of the seasonality of the work: Sod farms operate year-round, but for several months each year there is not much for strawberry pickers to do.

Santiago Flores knows all about that. Each summer, as the berry harvest wanes, he experiences the farmworker's version of vacation: a period of unemployment—some of it covered by state unemployment insurance, and some not—during which he can spend time with his kids, worry about paying the bills and rest up for the October start of the next strawberry season.

When that season commences, he works 10 hours a day, six days a week, for a wage that averages $8 an hour when you take into account both the base hourly wage of $6.75 and the bonus for each crate of fruit he fills.

In the meantime, his wife heads off each morning before dawn to a Santa Paula vegetable farm, where she works eight hours a day, five days a week, earning $8 an hour planting and picking organic tomatoes, squash and other crops.

It's not much, but it beats the wages in Mexico. The last time Guadalupe worked there, in 1976, she earned the equivalent of about $2.10 a day for a shift that started at 6 a.m. and ended at 5:30 p.m. At the time, the minimum wage in California was $2.50 an hour, according to the state's Industrial Welfare Commission.

It was the search for better wages that led the Flores family, like so many of California's estimated 1 million immigrant farmworkers, to enter the United States in the first place.

"People told me you could make a lot of money in California," Santiago says.

LIFE IN THE FIELDS

Santiago has been working in the fields since he was 16, when he left his family's subsistence farm in Oaxaca. He sought work first in the state of Sinaloa, where he picked tomatoes—and where he met Guadalupe in Culiacan, the state's largest city—and then La Paz on the Baja peninsula, where he picked cotton.

He crossed the border illegally in 1972 as a teenager and found work in San Diego County. For the next 14 years, he worked in the fields and lived under conditions that make even an overcrowded Oxnard apartment sound luxurious.

"We lived in the mountains, in caves," he says, "like gophers."

Mexican immigrants still live in the northern San Diego County hills, occupying cardboard shacks, caves and dwellings cobbled together out of plastic and other discards from the fields where they work. According to a 2004 report by the county's Regional Task Force on the Homeless, about 2,344 San Diego County farmworkers were officially homeless, living in such primitive encampments or in the fields themselves.

Although he was living and working in the United States most of the year, Santiago returned each fall to Mexico, where in 1975 he married Guadalupe. Santiago continued to work seasonally in California, eventually leaving San Diego County and moving north to Santa Maria, where he learned to pick strawberries. Guadalupe remained in Mexico, sometimes working but mostly tending the household and relying on the money Santiago sent back from the United States.

"When I was living in Mexico, I would just take care of the children," she says. "We had a house."

Santiago took advantage of the amnesty and citizenship provisions in the 1986 Immigration Reform and Control Act to become a legal U.S. resident. In 1993, he moved his family from Mexico to Santa Maria, where at first they lived in a motel. They moved to Washington to pick apples, strawberries and raspberries, to Sacramento to harvest vegetables, and then came back to Santa Maria. Nine years ago they moved to Oxnard and found work in the strawberry fields.

The one-bedroom apartment they shared before moving to their current home rented for $1,150 a month. Now they pay $858 for a four-bedroom apartment, and all the kids have their own beds. Guadalupe and Santiago sleep on a mattress on the floor because they're still saving money to buy a bed for themselves.

"As a parent, you have to take care of the children's needs first," Santiago says.

CHEAP FOOD

At $8 an hour, even if both Santiago and Guadalupe Flores worked full time all year, they would gross only $32,000 annually. But they do not work full time all year, so their income puts them well below the 2006 federal poverty level of $42,945 for a household of nine individuals, including four related children under 18.

It would not cost consumers much if farmworker pay increased enough to lift such families out of poverty.

According to the U.S. Department of Agriculture, farm labor accounts for about 6.5 percent of the price of fresh farm products. Double farmworkers' pay, and a dollar's worth of food would theoretically cost about $1.07.

According to the U.S. Bureau of Labor Statistics, the average American household spent $5,781 on food in 2004, the most recent year for which data are available. This included $3,347 for food consumed at home, and $2,434— 42 percent of the total—for food consumed at restaurants and other venues.

Of the total spent on food consumed at home, relatively little went for the kinds of products grown by Ventura County farmers. The average household spent just $561 on produce in 2004. And of that total, only $370 was for fresh fruits and vegetables; the rest was spent on processed produce.

Double farmworkers' pay, pass the increased cost along to consumers, and the average American household's produce bill would go up by $39.27 a year. Even if the wage increase raised food costs across the board by a similar amount— unlikely, since farm labor accounts for a much smaller percentage of the retail price for restaurant meals and processed foods—the average household would get dinged for another $404.67 a year. That's roughly eight bucks a week, the price of a movie ticket.

The flaw in this reasoning, according to growers and economic analysts, is that farmers cannot unilaterally pass along production cost increases. Prices for farm products are set by brokers, wholesalers and retailers, who increasingly deal in a global food marketplace and can tap cheaper sources in countries where labor, land and regulatory costs are lower than in the United States.

"Farmers are price takers, not price setters," says Charles Maxey, dean of the business school at California Lutheran University in Thousand Oaks, who has studied the county's agricultural industry.

The most likely effect if farm wages doubled overnight?

Picker's bags waiting to be emptied, Leavens Ranches orchard.

Gary Faye

Jessica Flores in the courtyard at Villa Cesar Chavez. Karen Quincy Loberg/Ventura County Star

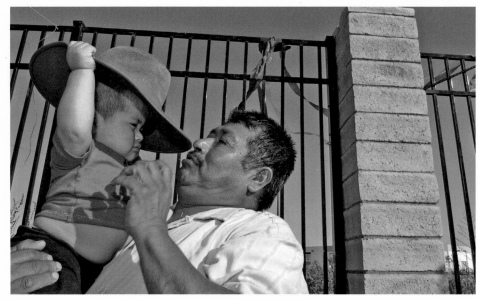

Santiago Flores and his grandson, Felix. Karen Quincy Loberg/Ventura County Star

"Farmworkers would make a lot of money for a short time until the farms start going out of business," Maxey says.

EYE ON THE FUTURE

Many elements of the Flores family's story are typical of California's farmworker population. The great majority, 73 percent, are men, and although 64 percent of the state's farmworkers are married, 28 percent live apart from their spouses. More than a quarter of those with children live apart from them.

This fragmentation of families—the men traveling to the United States to work, the women remaining in Mexico to raise children alone—is often cited by public-health workers as putting a profound strain on both individuals and on the social fabric of communities. And Santiago is not reluctant to talk about the emotional toll exacted by two decades in which he saw his wife and children only sporadically.

"It was very difficult for me to be by myself here," he says.

On this afternoon, the Flores household offers sufficient domestic tumult to reassure him that lonely solitude is no longer an option.

Guadalupe has just gotten home from a day spent planting cabbage on the farm where she works. Fatigue clouds her eyes, and her jeans and sweatshirt are caked with dirt. Some of the kids are already home from school, and are running in and out through the front door, accompanied by an ever-changing constellation of young friends and neighbors with backpacks and Barbie dolls and textbooks. The air is alive with chatter and footsteps and the banging of doors.

Santiago eventually slips in through the garage door, having picked up the other kids from school, and watches the activity quietly but with a contented smile. It is commonplace chaos, and all the more precious for its predictable, everyday nature.

Like generations of immigrants before, Santiago and Guadalupe think about the future in terms of the opportunities it will offer their children, not themselves.

"That's my goal, for them to study so they don't have to work in the field," Santiago says. "I want to get a better job, but I don't have a lot of education, a lot of English."

Evidence of their faith in the transforming power of education is on display in the Flores apartment. Although it is sparsely furnished—an aged console television, dinette set, sofa, chair and end table are pretty much it—the living area also features son Felix's graduation diploma from Hueneme High School, framed and granted a position of prominence atop the television, as well as daughter Janeth's certificate of promotion from middle school, similarly framed and displayed on the end table.

They're conscious of their good fortune in having traded a succession of motel rooms and overcrowded apartments for the relative serenity and security of Villa Cesar Chavez.

"It makes me really happy to be living here," Santiago says. "I'm not the only one who needs this type of living. We need housing for a lot of people."

 ## FARM LABOR BY THE NUMBERS

Estimated number of farm workers in Ventura County: 20,000

Number in California: 1.1 million

Percent born outside United States: 95

Percent born in Mexico: 91

Percent with U.S. citizenship: 10

Percent in country illegally: 57

Percent who have lived in U.S. more than 10 years: 47

Percent who have lived in U.S. less than two years: 18

Median age: 32

Percent who are male: 73

Average length of time employed in agriculture: 11 years

Percent employed year-round: 20

Percent who speak no English: 53

Median highest grade of schooling: 6th grade

Median hourly wage: $7.73

Percent who earn less than $15,000 a year: 75

Percent who earn less than $10,000 a year: 43

Percent with no health insurance: 70

Percent employed directly by growers: 63

Percent employed by labor contractors: 37

Number of farm labor contractors in Ventura County: 61

Number of farms: 2,318

Summer

Row-crop field ready for planting, Oxnard Plain.

Preceding page: Young citrus orchard, Santa Clara River Valley.

Gary Faye

CHAPTER FIVE
Designer Plants

It's a good idea to wear shoes you dislike when you visit Brokaw Nursery. To enter the fenced compound in Saticoy where thousands of young trees stand in regimental rows like skinny, inanimate soldiers, you must pass through a narrow gate with a roof over it. The floor of the passage is a shallow trough filled with a fine, gray powder, which clings persistently to footwear.

The powder is pulverized copper sulfate, and it is deadly to fungi. Fungi are the bane of California avocado growers, for whom Brokaw Nursery is the leading source of new trees. The main target of the copper powder is not shoe leather but an organism with the scientific name *Phytophthora cinnamomi,* a tongue twister that nearly everyone in the avocado business avoids by simply calling it "root rot."

A soil fungus that infects more than 1,000 types of plants, phytophthora causes avocado trees to drop their leaves, weaken and die. California growers first noticed its effects in the 1930s, and by the 1950s it had spread widely. Many experts initially feared it might doom the avocado industry.

It didn't. Over the past three decades, scientists and plant breeders at universities and nurseries have developed an array of weapons and tactics that have enabled growers to fight the fungus to a draw. The result is an uneasy peace predicated on unceasing vigilance, a sort of botanical Cold War minus the nukes. The key weapon in the farmers' arsenal has been development of rot-resistant avocado varieties, which means the fate of the modern avocado industry rests largely in the hands of plant propagators like those at Brokaw, who marry the hardiest root stocks to the best fruit-bearing wood.

But it is not only the avocado industry that depends on the research and development programs of nurseries, universities and labs. Every commercial crop grown in the county has its roots, so to speak, in the seeds and seedlings local farmers stick in the ground. And every seed or seedling is the result of a deliberate and painstaking process of selection intended to preserve and amplify desirable traits while minimizing or removing those that are undesirable.

That process produced Ojai pixie tangerines and Oxnard strawberries. It's behind the cornucopia of products sold by global seed giant Seminis, an Oxnard-based firm whose researchers have developed thousands of vegetable varieties with special customer-pleasing or pest-battling properties. In fact, it's behind just about every food product on your grocer's shelf.

Nature is the original plant breeder. Today, however, that role has been assumed by men and women in labs, manipulating the building blocks of life.

THE 'MOTHER TREE'

At Brokaw Nursery, the leading producer of avocado and citrus trees in California, the process that results in a tree suitable for transplant into a grower's orchard still starts with a seed. But after that, the process resembles nothing in nature.

Brokaw Nursery prefers to use the big seeds from avocados native to the West Indies because they are easier to work with, says Larry Rose, who manages sales and advises customers on plant cultivation for Brokaw. The seed is planted in a bed of soil, where it sprouts roots, stem and leaves, and develops into what plant breeders call a "nurse seedling."

Grafting a lemon tree, Brokaw Nursery, Saticoy.

Karen Quincy Loberg/Ventura County Star

A clipping from an adult tree that displays desired characteristics, such as resistance to root rot or ability to tolerate salty soil, is inserted into a slit in the trunk of the nurse seedling. The clipping sprouts its own roots, and then a clipping from an adult tree that produces the desired fruit variety is inserted into a slit in that newly rooted stem. The nurse seedling dies, and the new tree—marrying the root and fruit characteristics of two different varieties—is grown in a container at the nursery for another year before it's transplanted into an orchard.

Plant breeders are constantly looking for new varieties, testing their properties and hoping to find traits that will yield superior fruit or better resistance to pests and disease. It is a tedious and time-consuming process of trial and error that sounds a lot more exciting than it really is when Rose describes it.

"They crowd a bunch of trees close together, force them to have sex, and then take the seeds from that and plant them out," Rose says.

Although there are several commonly used root stocks, nearly all the fruit-producing wood used by Brokaw and other nurseries is of the Hass variety, which now accounts for 95 percent of all commercial avocado plantings in California and 80 percent of the global harvest. And every one of those trees is descended from a single "mother tree," a fluke discovery by La Habra postal carrier Rudolph Hass, who in 1926 found a tree producing odd purple-black fruit growing in his yard among a group of avocado seedlings he'd bought from a nursery.

No one knows the variety of seed that produced that tree, but Hass' children proclaimed its fruit far superior to the green-skinned Fuerte variety then dominating the industry. Hass patented the new variety in 1935, named it after himself, and contracted with Whittier nurseryman Harold Brokaw to grow and promote it. Brokaw did so with spectacular success, helped by the Hass variety's characteristics: higher yield, better durability to withstand shipping, and a richer and nuttier taste.

Brokaw's nephew Hank, who with his wife, Ellen, owns Saticoy-based Brokaw Nursery, tended the original mother tree for many years, but in 2002 it succumbed to the dreaded root rot. The tree was cut down, and its remains were tucked away at the Saticoy nursery to await disposition appropriate to its status as founder of an industry.

GENERATIONS OF DAUGHTERS

Like avocados propagated using the Brokaw technique, the strawberries that growers transplant each fall into their fields on the Oxnard Plain are the result of a long and complex process that takes years to yield results. It typically begins more than 500 miles from coastal Southern California, in a small building behind an equipment yard on the dusty outskirts of Redding.

That's where Lassen Canyon Nursery lab manager Pete Stone places a strawberry plant obtained from a breeder—typically the University of California, which has developed most of the major commercial varieties grown in the state—under the lens of a microscope. As he peers through the eyepieces, Stone uses a tiny scoop to remove a chunk of tissue no bigger than a millimeter in width from the plant's meristem, a small region of actively dividing plant cells located just above the roots.

Meristem cells are the plant-world version of human stem cells, capable of developing into any of the many specialized cells that serve different functions in an organism. Just as stem cells can become brain cells, red blood cells or skin cells, meristem cells can become leaves, flowers, or roots. And because they are isolated from the plant's vascular system, meristem cells generally avoid infection by any of the myriad viruses and other pathogens found in soil, air and water, Stone says.

In his lab, Stone places the tiny lump of meristem tissue into a small test tube containing a gel-like growth medium. Within weeks, the cells have developed into a complete strawberry plant about two inches long.

The development of leaves, stems and roots from a nearly invisible blob of undifferentiated cellular materials seems a bit like alchemy even to those familiar with the process.

"It still amazes me, and I'm the one that does it all the time," Stone says.

Once they have developed sufficiently, the tiny plants are removed from the test tubes, transplanted into 4-inch pots and moved to a special greenhouse, where they are bathed regularly in mist and allowed to develop into full-sized plants. Those plants are then transplanted into big plastic bins filled with sterilized soil in a large screen-walled greenhouse. Each plant sends out numerous runners; each runner develops several tiny plantlets, which take root in the sterile soil.

This next generation of daughter plants—genetically identical clones of their mothers, which in turn were genetically identical clones of the original plant that provided the meristem cells—is harvested and transplanted into the nursery's foundation beds.

Trimmed strawberry plant, ready for transplant into field. Karen Quincy Loberg/Ventura County Star

The process is repeated. And repeated. Each step—multiplying a mother plant into numerous daughter plants, and then transplanting those offspring to produce another generation of daughters—requires an entire growing season. By the time Lassen Canyon Nursery has turned a single source plant into a commercially useful quantity of seedlings ready for growers to transplant into their fields, Stone says, six years will have elapsed.

BOTANICAL TRICKERY

The strawberry amplification process seems protracted, but it's the most effective way to turn a small number of disease-free plants into the vast quantity needed by commercial berry producers. A single matriarch, the breeder-supplied plant that found itself under Stone's microscope in the meristem lab, has become

1.5 million plants by the time its offspring are in the nursery's production fields. Each of those plants generates 30 or so seedlings for sale.

And growers need a lot of plants. Oxnard Plain farmers typically plant between 23,500 and 26,000 plants per acre when they establish their fields, according to grower Cecil Martinez. Ventura County growers cultivated nearly 12,000 acres of berries in 2006, which means nurseries needed to churn out 300 million seedlings just to meet local demand.

Lassen is one of the leading producers of plants for California's commercial strawberry growers, sending about 230 million plants each year to Oxnard, Santa Maria, Salinas and Watsonville.

The plants that end up in Ventura County are raised in fields near Macdoel, a blink-and-you-miss-it town near the California-Oregon border on the edge of the Cascade range. Those fields have a scenic view of snowcapped Mt. Shasta, and are at a relatively high elevation of more than 4,000 feet.

That high elevation is crucial. It gets cold there pretty early in the fall, and when the cool weather arrives, it signals the plants to prepare for dormancy, which they do by frantically storing energy in their roots and crowns.

In September and October, after the plants have experienced about 240 cumulative hours of temperatures below 45 degrees, they're harvested by machine and hauled to the nursery's trim sheds, where hundreds of workers gather before dawn each day during the harvest. They use heavy steel blades to trim the final crop of daughter plants, cutting off all but an inch of so of the stems and leaving about six inches of roots. The trimmed plants are packed in cartons, refrigerated and trucked almost immediately to customers in California's berry growing regions.

When each of those chilled plants is set back in the warm ground 48 hours later and bathed by Southern California's toasty autumn sunshine, it draws upon the energy it was tricked into storing for dormancy and explodes back into life.

"It thinks it's on the beach in Honolulu or something," Stone says.

Evaluating bean seeds for vigor, Seminis, Oxnard.   *Karen Quincy Loberg/Ventura County Star*

The dormancy trick gives the plant a metabolic jump start, and it grows and sets fruit more rapidly than it otherwise would. Oxnard Plain growers can be harvesting strawberries 100 days after sticking one of Lassen's seedlings in the ground. That means they can begin selling fruit in the dead of winter, when it commands a premium price.

GLOBAL REACH

Compared to the elaborate reproductive process that generates commercial avocado and strawberry plantings, you might expect the story behind the humble vegetable seed to be a model of simplicity.

You would be wrong.

Even this most elemental of agricultural techniques—saving the seeds from one

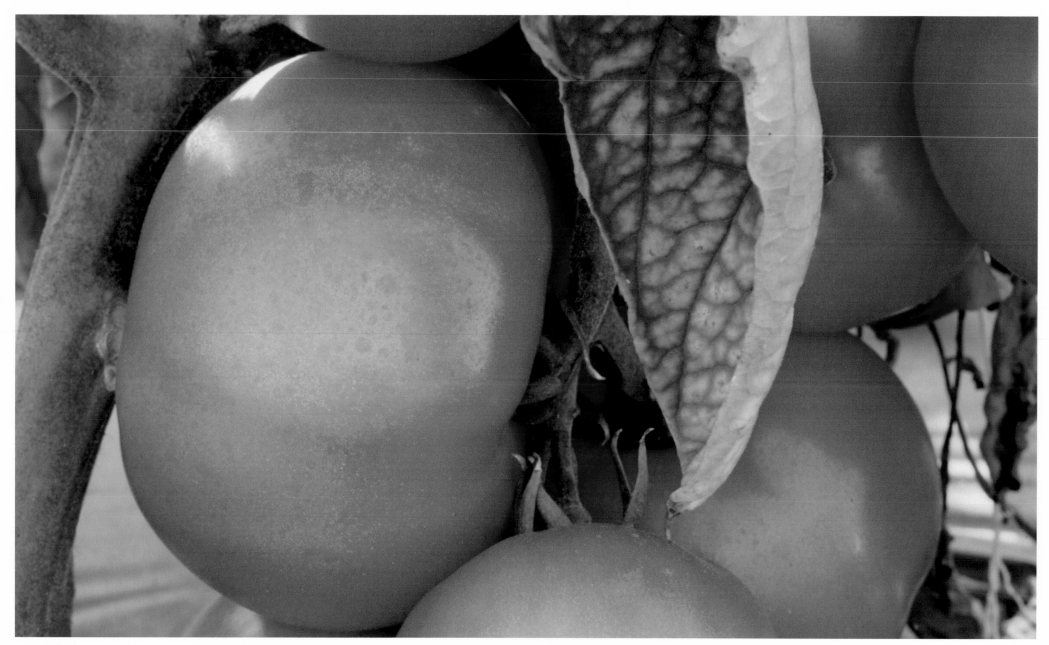

Tomatoes ready to harvest, Oxnard Plain.

Gary Faye

year's crop to plant the next, a practice as old as farming itself—has undergone a radical transformation to meet the demands of modern food production and marketing. Just how complicated the modern seed story has become is made clear by a visit to the Oxnard headquarters of Seminis Inc., the world's leading developer, grower and marketer of vegetable and fruit seeds.

Seeds from the company's far-flung production fields (it produces seeds in 23 countries) eventually make their way to the Oxnard plant, which occupies a nondescript industrial park on the eastern edge of Oxnard. The plant houses company headquarters along with a refrigerated warehouse, milling and drying facilities, labs and greenhouses, and is the largest of the company's 11 U.S. facilities.

The milling department, where seeds are sorted by size before being packaged, occupies most of the floor area. Sorting involves a lot of shaking, as the seeds drop from overhead chutes onto vibrating platforms fitted with screens of various apertures. The precision-planting machines used by today's growers are calibrated for seeds of a particular size, so Seminis tries to ensure uniformity in each canister of seed.

Upstairs in the main building is a warren of laboratories, many looking like overgrown high school science classrooms, where employees analyze everything from the DNA of individual plants to the root length and germination success of different seed varieties. Seminis says it spends $50 million a year on research.

Much of that research is devoted to tailoring products to meet the complex demands of growers in a ferociously competitive industry. For example, the company sells lettuce seeds customized to perform at their best under the average temperatures and hours of daylight found at specific places and at specific times of the year. As a farmer seeking to maximize yield and minimize uncertainty, you would not plant the same variety in the Salinas Valley that you would in the Imperial Valley, and the seeds you plant today are a subtly different strain than the ones you will plant a month from now, even though they may well produce superficially identical kinds of lettuce.

The company has also developed varieties customized to match consumer preferences: "personal size" watermelons; a boat-shaped lettuce leaf for carb-shunning diners who disdain bread and tortillas but don't want to give up sandwiches and tacos; "heat-free" jalapeno peppers.

Seminis (the name means "seed" in Latin) is a corporate hybrid, assembled through the merger of four older seed companies. One of those was Saticoy-based Petoseed, founded locally in 1950. The firm now commands more than a fifth of the global market for commercial fruit and vegetable seeds, and was purchased in 2005 year by Monsanto, a biotech and chemical giant based in St. Louis. Monsanto is a world leader in the use of sophisticated genetic modification techniques to engineer chemical resistance and other traits in its crops—soybeans and corn, for example, that are immune to the company's bestselling weed-killer, Roundup.

That technology has touched off a firestorm of controversy through the world, with critics warning that the spread of genetically modified organisms, or GMOs, represents a giant uncontrolled experiment that could imperil human health and the environment. Defenders of the technology argue that it merely represents an advanced form of the same kind of manipulation that plant breeders have employed for hundreds of years.

Seminis has so far sidestepped the controversy by continuing to produce its seed varieties using traditional hybridization and pollination techniques, although it has begun using a DNA analysis technique pioneered by Monsanto to speed up that process by identifying which of many potential cross-breeding candidates actually possess the genes that code for desired crop traits.

Those traditional techniques are themselves complicated and often tedious, involving the meticulous crossbreeding of thousands of different plants by transferring pollen from one to another—sometime using tweezers—and growing out the progeny to see what happens. It can take a decade or more to produce a new variety.

Even generating a crop of seeds for commercial sale from well-established plant lines takes time and a great deal of labor.

Pollinating tomato blossoms by hand.

Gary Faye

Pollinating tomatoes.

DESIGNER PLANTS

No vegetable is more homey and humble than the zucchini; even the most inept home gardener can usually manage to coax a plant into producing squash. But to generate commercial quantities of the hybrid seed responsible for that plant involves several steps.

Here's what it looks like in a 30-acre Seminis production field near Camarillo.

Rows of two parent squash varieties are planted, one type being designated the female and the other the male. The females, which will develop offspring blending the desired traits of both parents, outnumber the males six rows to one, and are sprayed with a chemical that ensures they produce only female flowers. The late-blooming males are planted two weeks earlier than the females, to make sure both varieties bear flowers at the same time.

"You can lose a whole crop if your male flower is too early or too late," says Dan Holmes, director of production logistics for Seminis.

Beehives are set into the field, and the insects carry pollen from the male plants to the female plants. Periodically, workers make their way through the fields, plucking off any male flowers that somehow sprout on the female plants, to make sure those plants don't pollinate themselves. Self-pollination would produce nonhybrid seeds, contaminating the entire crop.

The male plants are then destroyed. When mature, the squash on the female plants are mechanically harvested and pulverized, to liberate the seeds from the pulp, which is washed away. (Other types of vegetables produce "dry" seeds that don't require such elaborate processing.)

Some plant varieties do not lend themselves to open-field pollination and require human intervention: Workers must walk the rows, moving pollen from one flower to another by hand. For some hybrid tomato varieties, Holmes says, it can take 5,000 to 8,000 hand pollinations to produce a pound of seeds. Not surprisingly, such seeds are expensive, running as high as $1 per seed.

Seminis also produces seeds for the home gardener, and many of these are

Pole tomato field, Oxnard Plain.

Gary Faye

tested in a trial plot in Saticoy. The field is an amateur horticulturist's dream, row after perfect row of lush tomatoes, cucumbers, beans, corn, lettuce. Some of the plants are being grown from seeds plucked off the Seminis production line, to make sure they perform as advertised. Some are competitors' products, being compared to Seminis varieties.

But many are potential candidates for addition to the Seminis product line. The company scours the world looking for new plant types, which it evaluates for desirable traits. And it is here that the elaborate architecture of modern plant breeding and selection, so focused on the competitive interests of the commercial food production and marketing industries, returns to its more elemental roots.

"We're looking for variety, flavor, earliness and attractiveness," says Dan Croker, regional sales manager for the Seminis garden division. "We want beauty in our fruit."

Tomato packinghouse, Camarillo.

Gary Faye

CHAPTER SIX
Going Global

With a loud whine, the Lady Korcula's deck crane heaves a heavy steel cage full of cardboard boxes high into the air, swings it over the ship's rail and drops it into the gaping mouth of the cargo hold. Workers there quickly unload and stow the boxes, and the crane swings back to hoist another load off the dock. The process is repeated again and again with nearly metronomic precision, each load landing with a clangorous thud in the belly of the 510-foot freighter.

The Lady Korcula has been moored at the Port of Hueneme for four days, and dockworkers are packing its refrigerated hold with Southern California citrus. Much of the cargo consists of lemons that just a few weeks ago were dangling from trees in Santa Paula, Saticoy, Moorpark, Ventura. In about 12 days they'll be in Japan, where dockworkers will re-enact this industrial dance in reverse.

There's money to be made sending the products of local farms and ranches abroad. And for many growers, exporting is not just a potentially profitable option, it's a necessity. Capitalizing on the area's moderate climate and fertile soil, Ventura County farmers produce far more food than could possibly be consumed locally. Without access to customers in distant markets, they would inundate their neighbors with a flood of fruits and vegetables.

The annual lemon consumption of Ventura County's 800,000 residents, for example, could theoretically be satisfied by about 170 acres of trees. Yet the county actually has more than 22,000 acres of lemon orchards.

But Ventura County farmers are increasingly finding that the global marketplace offers pitfalls as well as opportunities. In a world where nearly any farmer can reach nearly any customer within a few days, growers in regions with relatively high land, labor and production costs—such as coastal Southern California—often find themselves at a competitive disadvantage.

And it's a world that grows smaller and more integrated by the week. Evidence floats dockside on this fine May day at the Port of Hueneme, an international intersection where local farming meets the global economy.

The Lady Korcula is a Swedish-chartered freighter, built and crewed by Croatians, flying the flag of the Marshall Islands. It carried bananas from Ecuador to California before picking up its load of West Coast citrus; when it leaves Japan, it will be full of used cars bound for Peru. There, it might pick up frozen fish before returning to Ecuador to take on another load of bananas bound for California and Japan.

DRIVEN BY DEMAND

In much of the United States, farming began as a subsistence activity and only gradually became a commercial enterprise. Settlers initially planted crops to feed themselves and to fuel the farm machinery of their time: horses, mules and oxen, which were powered by hay and grain rather than the petroleum gobbled by modern tractors.

Once they'd established self-sufficiency in basic staples, farmers with access to adequate land, investment capital and labor could raise surplus crops and livestock for cash sale in the marketplace. In much of the United States, this process of transformation from homestead to commercial enterprise took decades.

Not so in California.

Sorting lemons at Saticoy Lemon Association packinghouse.

Gary Faye

"California agriculture has always been demand driven," researchers Warren Johnston and Alex McCalla wrote in a 2004 report for the University of California's Giannini Foundation of Agricultural Economics. "It was never subsistence, family-farm agriculture like that which characterized much of early U.S. agriculture; rather, it was driven by entrepreneurs seeking riches by serving high-value and/or newly emerging markets. These markets were generally distant and often foreign: hides and tallow to the United Kingdom and Boston; wheat to Europe and beyond; fruits, nuts and vegetables to the East Coast, Europe and, more recently, Asia; and wine to the world."

That's almost a perfect description of Ventura County's agricultural history. Growers here have always relied on access to distant markets. And access to markets has always required access to transportation.

Ships came first. As early as the 1820s, rancho owners were dispatching board-stiff bundles of cattle hide to New England shoemakers via sailing vessels that sent small boats to shore through the surf.

The process became more convenient in 1871, when a wharf was constructed at the town of Hueneme, taking advantage of a deep underwater canyon that allowed ships to approach close to shore without running aground. Ventura completed its own wharf a year later. Both projects were spearheaded by growers seeking a way to move grain from their Ventura County fields to customers in urban markets.

But ocean shipment in the late 19th and early 20th centuries was not suitable for perishable products. Before the Panama Canal was completed in 1914, a sea journey from California to the East Coast took four months or longer.

Citrus and vegetable growers didn't have long to wait for a more suitable transportation system. The transcontinental railroad line knit the nation's east and west coasts together in 1869, and tracks reached Ventura County in 1886, when Southern Pacific extended a branch line down the Santa Clara River valley from Newhall. The first train reached Ventura in 1887, setting off a real-estate boom and providing the fledgling citrus industry a means by which to send lemons and oranges to the rest of the country.

The global marketplace, however, can be a treacherous place. Ventura County wheat and barley growers first discovered this in the late 1880s when skyrocketing production elsewhere in the United States caused grain prices to collapse, and increasing production in Canada and Russia displaced American grain from the European market.

Like their predecessors in the grain business more than a century ago, many of today's Ventura County growers remain economically dependent on distant markets and economically vulnerable to distant competitors.

"We used to figure a third of our market would be export to Japan," says Link Leavens. "We don't figure that anymore."

A SHRINKING MARKET

Perhaps no local crop illustrates the promises and perils of the global marketplace better than lemons, which for more than half a century were the most valuable agricultural commodity produced in Ventura County. They still account for more of the county's irrigated acreage than any other crop, although that figure is dwindling.

The United States ranks fifth in the world in lemon production, according to the U.S. Department of Agriculture's Economic Research Service, behind India, Argentina, Spain and Iran. Virtually all American lemon production takes place in California and Arizona, with a minor contribution from Florida. California is by far the dominant producer, accounting for nearly 90 percent of the domestic crop.

With its nearly frost-free coastal climate, Ventura County is perfectly suited to production of lemons, a subtropical crop that cannot tolerate temperatures much below 30 degrees, and its growers produce nearly half the state's total. Ventura County orchards produced 794 million pounds of lemons in 2005, which is enough to satisfy the average annual demand of 113 million Americans, according to federal consumption data.

A significant share of U.S. lemon production historically has been destined for export, mainly to Asia, where the fruit is a popular accompaniment to seafood and tea. But American lemon exports have been dwindling as production has expanded in other exporting countries, such as Argentina, where it's cheaper to grow fruit.

Argentine growers, who last year produced about 1.3 million tons of lemons and exported 390,000 tons, have emerged as major competitors for U.S. growers in the global marketplace. Lemon acreage also has increased in Mexico, South Africa and Spain.

Between 2001 and 2005, annual U.S. exports of fresh lemons—those that bring the highest price and the greatest profits for growers—dropped from about 125,000 tons to 114,000 tons, according to the USDA's Foreign Agricultural Service.

Shipments to Canada, Australia and China rose, but they were more than offset by the decline of exports to the top foreign destination for American lemons: Japan, where sales of American lemons and limes fell 23 percent between 2001 and 2005, from 76,000 tons to 59,000 tons.

That drop in sales to Japan has directly affected Ventura County growers. Most are members of the Sunkist marketing cooperative, which ships Southern California citrus to Japan using chartered vessels—such as the Lady Korcula—that dock at the Port of Hueneme. Those ships are operated by NYKLauritzenCool USA Inc., headquartered in Sweden, which operates a refrigerated warehouse complex at the port.

At one time, the company was hauling 8.5 million cartons of Sunkist fruit to Japan a year. That's now down to about 5 million cartons, says Mike Karmelich, who runs the shipping company's operations at the Port of Hueneme. Other producing nations have moved into the Japan market.

"It's become much more competitive because of the global economy," Karmelich says.

DWINDLING ACREAGE

The growth in global lemon production has enabled growers in other countries to make inroads into the U.S. market as well. Since the mid-1970s, when there were virtually no imported fresh lemons to be found in the United States, imports have grown to command about nine percent of domestic sales, according to the USDA.

Even small changes in the local share of the domestic and export markets for fresh lemons can have a disproportionate effect on the profitability of lemon production in a high-cost region such as Ventura County.

Forty to 50 percent of each year's lemon production in Ventura County is sold for juice, concentrate and other products, at a price that doesn't cover the cost of growing, picking and packing it. About half the crop is sold on the fresh market and earns growers a profit. Perhaps 15 percent of the total crop is top grade and brings the highest price.

With nearly half the crop already generating no real profit, growers can ill afford to lose buyers for what remains. As a result, the shrinking export market has had a visible effect on the local landscape.

Between 1998 and 2005, land planted in lemon trees in Ventura County dropped from an all-time high of 27,707 acres to 20,875 acres, according to the Agricultural Commissioner's Office. Many of the bulldozed orchards were replaced by more lucrative strawberries, which accounted for 5,557 more county acres in 2005 than in 1998.

It is no coincidence that 1999 was the year strawberries became the county's top crop in total value, the first time since 1947 that lemons had not occupied the No. 1 spot.

Lemons popped back to the top in 2000, but slid back again in 2001. They now rank third in value, behind strawberries and nursery stock and just barely ahead of avocados, which are closing the gap.

Many growers expect Ventura County lemon acreage to shrink further in the next few years. To be honest, they don't regard that as a bad thing. Fewer lemon orchards means fewer lemons, which might bring supply more closely in alignment with demand and stabilize prices.

"Five or six thousand acres out of the ground helps," Leavens says.

EXPORTS AT A PRICE

The modern Port of Hueneme has little in common with its 19th century predecessor. In place of the 900-foot wooden wharf built by Thomas Bard's Hueneme Wharf and Lighter Co. in 1871, the port today consists of a 2,300-foot-long channel leading to a dredged harbor 35 feet deep, shared by the commercial operator and the Navy.

The commercial port has two concrete wharves capable of accommodating deep-draft vessels. The north wharf is 1,450 feet long and provides two berths, used mainly by ships carrying new cars. The south wharf is 1,800 feet long and provides three 600-foot berths. Those berths, used mainly by fruit carriers, are adjacent to three refrigerated sheds that together have 39 truck bays and a collective storage area of 164,651 square feet.

On this sunny May day, the Lady Korcula is moored at the south wharf, and its deck crane is the central character in an industrial ballet.

The fruit boxes, decorated with logos and script in vivid hues, are packed into a yellow cage gripped by a bright red hook dangling from the crane's long gray arm. As all that colorful metal and cardboard swings overhead against the deep blue sky, headed for the hold below deck, it's easy to become mesmerized by the spectacle.

Or it would be, if not for the forklifts buzzing like irritated wasps in and out of the nearby warehouse. Ignore events on the ground while watching fruit in the air and you risk being speared or squashed.

The cargo being loaded on this day consists of 130,000 cartons of fruit, about half of which are lemons. Each carton weighs nearly 40 pounds, meaning the freighter will be carrying about 5 million pounds of California citrus across the Pacific.

Boarding the Lady Korcula is like entering a small European village. Capt. Dean

Sunkist fruit being loaded aboard Japan-bound freighter. Karen Quincy Loberg/Ventura County Star

Botica is Croatian, as is the rest of the 21-man crew, and the warning signs posted around the ship are in English and Croatian. The crew members speak heavily accented English, when they speak English at all, and they treat an unexpected group of visitors with unfailing politeness. The crewmen are all from the same region of Croatia, as is the ship, which is named after an island in the Adriatic Sea.

From the Port of Hueneme, the ship will traverse what Botica calls the "great circle" route: up the West Coast and then along a swooping arc that takes the ship into the northern Pacific before swinging south toward Japan.

In winter, bad weather makes that route nearly impassable, Botica says. Spray freezes on everything above the waterline, 45-foot seas sweeping over the deck

Cabbage field and tractor, Oxnard Plain.

Gary Faye

dislodge cargo and toss the vessel around like a bathtub toy, winds of 50 to 60 knots push the ship off course and slow its progress.

At this time of year, though, conditions are a bit different. Botica smiles broadly when asked what the journey is like in May.

"Beautiful," he says.

Even when the ocean is flat and the air is still, moving Ventura County fruit halfway around the world exacts a price. Botica consults a chart and determines that at its cruising speed of about 19 knots, the Lady Korcula burns roughly 45 metric tons of bunker fuel a day.

That fuel—a tarry refining residue that must be warmed before it's thin enough to pump, and which is reviled by environmentalists for the quantity of sulfur dioxide, nitrogen oxide and other pollutants it produces when burned—costs about $360 a ton, Karmelich says. A year ago, that same fuel cost $150 a ton, he says, but bunker fuel prices are following the same ascending trajectory as the prices of gasoline, diesel and other petroleum products.

To deliver 5 million pounds of Southern California citrus to Japan, the Lady Korcula will burn more than 1.1 million pounds of fossil fuel at a cost of $8.4 million.

SELLING DIRECT

Not everyone believes global marketing is the best path to profitability.

At 6 a.m. each Sunday, Jim Churchill and Lisa Brenneis leave their Ojai home, squeeze into a Chrysler Town and Country minivan stuffed with crates of fruit, folding tables, baskets, signs and other tools of the grocer's trade, and then head for Hollywood at a speed that invites attention from the California Highway Patrol.

When they reach Hollywood, they head for the site of the neighborhood's legendary farmers market, which each Sunday morning draws throngs of shoppers, and as many as 100 growers and 50 other vendors, to several crowded blocks of Ivar and Selma streets. By 7:15, they're using the van's contents to erect

Jim Churchill setting up at Hollywood Farmers Market.   *Karen Quincy Loberg/Ventura County Star*

an open-air grocery store. By 8, they're selling pieces of fruit from their 17-acre citrus and avocado orchard directly to consumers.

Selling your own crop directly to buyers is a time-consuming activity. Most farmers say they find it sufficiently challenging to cope with weather, pests, labor contractors and balky tractors, and display no eagerness to add negotiations with market managers and demanding shoppers to their workload. Instead, they rely on marketing cooperatives such as Sunkist, grow under contract for big shipping and marketing companies such as Calavo and Sunrise Growers, or sell their fruit and vegetables to brokers or other intermediaries.

But for the relatively small number of growers who do sell directly to their customers, the choice is based on compelling logic.

Roadside produce stand, Somis.   *John Krist*

"We sell direct because we want to keep the money," says Churchill, who also sells fruit each week at markets in Ojai and Santa Barbara.

Keeping the money is not easy for most growers. As food travels the long and increasingly convoluted path from farm to dinner table, a portion of every consumer dollar is siphoned off at each stage, leaving a steadily shrinking share of it for the farmer who conjured that food out of sunshine, soil and water. In 1952, 47 cents of every dollar Americans spent on food made its way back to the farmer who produced that food. Today, the farmer's share of that dollar is 19 cents.

Processing and marketing now account for 81 cents of every dollar Americans spend on food, according to the U.S. Department of Agriculture. Advertising alone takes 4 pennies out of every 100—more than the typical farmer keeps as profit. Additional bites are taken by brokers, shippers, processors, distributors and retailers.

In response, growers have three basic options. They can switch crops, abandoning those with narrowing profit margins for others that offer greater returns. They can try to compensate for the shrinking profit on each pound of produce by growing more of it, making up in volume what they're losing on margin. Or they can try to bypass the long food chain altogether, selling their products directly to consumers.

Ventura County growers have adopted all three strategies.

PAYING FOR CONVENIENCE

In 2004, the average American household spent $5,781 on food. This included $3,347 for food consumed at home, and $2,434—42 percent of the total—for food consumed at restaurants and other venues, according to the U.S. Bureau of Labor Statistics.

The growing American appetite for convenience, whether that means a fast-food burger and fries or pre-washed salad served out of a bag at home, has meant a growing share of the consumer food dollar is spent on processing. As a rule, the less a food item looks like a fresh-from-the-field vegetable or piece of fruit, the smaller the farmer's share of the purchase price, according to agricultural economists.

For example: Stewart Smith, a University of Maine professor in resource economics and policy who studies changes in the structure of the nation's food system, has calculated that when you buy a bag of potato chips, about 2.5 percent of the purchase price returns to the farmer who grew the potatoes. When you buy a can of potato "crisps" such as Pringles, on the other hand, only 0.5 percent of the purchase price returns to the grower. The difference is captured by the company that mashed, extruded, molded and baked the potato starch into a snack food.

Potatoes, it is worth noting, are the most commonly consumed vegetable in the United States. Americans consume about 129 pounds per person per year, according to the USDA, two-thirds of them in the form of fries, chips and other processed products. Tomatoes rank No. 2, at 94 pounds per person—again, the vast majority in processed form. It's a long way back to No. 3, lettuce, at 33 pounds per person per year.

But even when consumers buy fresh fruits and vegetables that have not been sliced, diced, mashed, extruded or otherwise transformed by processing, the gap between what they pay and what the farmer gets paid can vary substantially depending on where the purchase occurs.

MAKING THE CONNECTION

For example, Churchill was selling his pixie tangerines at the farmers markets in early summer of 2006 for $1.50 a pound. The markets charge a vendor fee of 6 to 8 percent of gross sales, but otherwise Churchill gets to keep the money he collects.

The same week, Ojai Valley pixies were selling for $3.99 per pound at a Vons market in Ventura. The higher price reflected additional cost-adding steps in the food chain: The pixies at Vons had moved from Churchill's orchard—and the orchards of other members of the Ojai Pixie Growers Association—to a wholesaler, Melissa's/World Variety Produce. Melissa's then sold the fruit to the retailer.

The wholesaler and retailer each took a profit and marked up the price to recover the costs of advertising, transportation and labor. Churchill and his pixie-growing colleagues actually received less per pound for tangerines sold through that route than they did from farmers market sales, closer to $1 per pound than $1.50. The compensating advantage is that they were able to move a lot more fruit.

Churchill calculates that in 2006 he sold about 99,000 pounds of pixies to wholesalers and 42,000 pounds directly to his retail customers, which include Monterey Market and Berkeley Bowl Marketplace, both in the San Francisco Bay Area. He sold another 11,000 pounds of pixies through the Hollywood, Santa Barbara and Ojai farmers markets.

"When we started with pixies, no one would sell them. So we started (direct sales) out of necessity," Churchill says. "We're really creating a market where no market existed."

There's another reason to sell through face-to-face transaction with consumers, according to Churchill.

"I would say we also do farmers markets for fun," he says, describing the interaction with customers as a chance to teach people about food and farming, and to re-establish the consumer-grower connection that has largely been eroded by modern food marketing and retailing practices.

The combination of economic and social advantages—coupled with a growing consumer interest in healthy, fresh food—has meant an explosion in direct-to-consumer marketing by growers. According to the USDA, the number of farmers markets in the United States jumped 111 percent between 1994 and 2004, and there are now more than 3,700 of them.

More than 19,000 farmers nationwide report selling their products only at such markets. Throw in pick-your-own operations, and growers that sell online or through other channels, and the number of farms involved in direct sales rises to more than 93,000.

Even though it is growing, that number represents but a tiny fraction—4.4 percent—of the nation's 2.1 million farms.

Fruit display, Ojai Farmers Market.   John Krist

WHERE YOUR FOOD DOLLAR GOES

Processing and marketing:	81¢	Farmer	19¢
Labor:	38.5¢	Labor:	6.5¢
Packaging:	8¢	Water:	2.5¢
Fuel and electricity:	3.5¢	Transportation:	1¢
Transportation:	4¢	Pest control:	1¢
Advertising:	4¢	Fertilizer:	1¢
Depreciation:	3.5¢	Fuel and electricity:	1¢
Rent:	4¢	Taxes:	0.5¢
Taxes:	3.5¢	Equipment, repairs, maintenance:	0.5¢
Interest:	2.5¢	Weed control:	0.5¢
Repairs:	1.5¢	Insurance:	0.5¢
Other:	3¢	Other:	1¢
Corporate profit:	**4.5¢**	**Grower profit:**	**3¢**

Note: Numbers do not add up to 100 because of rounding. Figures reflect a national average, based on retail purchase of domestically produced farm foods.

Autumn

Pumpkins and sunflowers, Faulkner Farm, Santa Paula.

Preceding page: Sunflower, Faulkner Farm, Santa Paula.

Gary Faye

Chapter Seven

A Season of Worry

American farmers produce more than a billion pounds of pumpkins each year, and although most of the bright orange gourds wind up decorating front porches on Halloween, the bulk of the remaining crop end up as pie filling.

And that helps explain why Cecil Martinez seems a bit glum as he sits in his compact SUV alongside Gonzales Road in Oxnard, watching a crew pick strawberries less than two weeks before Thanksgiving of 2006.

"We're producing too much," Martinez says.

Fresh strawberries are, in the natural order of things, a spring and early summer crop. But plant breeders have developed varieties that can produce ripe fruit in the abbreviated daylight of autumn. Because Ventura County is one of the few places warm enough to grow delicate berries outdoors at a time of year when folks in most of the country are pulling winter coats out of mothballs, growers have an opportunity: The lack of competition from other regions means Oxnard Plain strawberry producers can pretty much have the autumn market to themselves, typically a prescription for high prices and fat profits.

Chasing those profits, local growers have more than doubled the amount of land devoted to fall berries just since 2002. Although most of the county's strawberry acreage continues to be planted for winter and spring harvest, more than 3,700 acres are now devoted to fall harvest in Ventura County, according to the California Strawberry Commission.

American consumers, however, have proved reluctant to trade their pumpkin pies for strawberry shortcake during the final months of the year, which remain the domain of roots, nuts and winter squash. Odds are that any berries on the menu for Thanksgiving dinner are cranberries. Fall strawberry production has expanded faster than consumer demand, cutting into prices and profits.

Autumn may be harvest time for some Ventura County strawberry growers, but for other farmers and ranchers, it is a time of waiting and preparing for next year.

It also is a time for worrying. If you spend much time around farmers, you quickly learn that they worry a lot.

WELL-TRAVELED PESTS

Link Leavens is worrying about a bug. It's tiny, about an eighth of an inch long as an adult and even smaller as a nymph. It's called the Asian citrus psyllid, and it feeds on lemon, grapefruit and other citrus trees, the tiny nymphs burrowing into young leaves and shoots and slurping up plant juices.

The insect's feeding habits damage trees, but that's not the main reason growers fear it. The bug transmits another pest, a bacteria responsible for greening disease, which University of California researchers have called "one of the most devastating diseases of citrus in the world."

Citrus greening disease stunts tree growth and causes fruit to drop or prevents it from ripening. It first showed up in the United States in 2005, affecting orchards in Florida. Pest experts warn that the psyllid that transmits greening disease may soon make its way to California from Florida, Texas or Mexico, where it has already established a presence, potentially bringing the devastating disease with it.

As if to serve warning, a similar plant pest from Florida—the citrus leafminer—showed up in local orchards in 2006 and within months had spread across the entire county, Leavens says. So far, the leafminer is a nuisance not a calamity, but it's a reminder—as if Leavens or any other grower really needed one—that farming often seems like one long, inconclusive battle with an endless succession of adversaries.

Weather is another of those potential adversaries.

"We're waiting for rain," Leavens says on a hot, dry day in mid-November. The year's lemon and avocado harvests are completed, but the trees are already devoting energy to producing next year's fruit, and that requires water.

The weather this November has been hot, marked by a succession of dry, 80-degree days. For some county residents, that means an opportunity to stage photos on the beach that will become envy-inducing Christmas cards mailed to friend and relatives in colder climes. For farmers and ranchers, however, it means running irrigation systems. That costs money, and growers are waiting impatiently for nature to step in and begin paying the bills.

Hot weather is often accompanied by Santa Ana winds, which can damage trees and scar fruit if they gust hard enough. Every red-flag alert is as worrisome to orchard managers as it is to firefighters.

REPORT CARDS DUE

But if natural wind can be a powerful foe, artificial wind can be a grower's best friend.

As autumn arrives, workers begin testing the wind machines that protect Leavens Ranches orchards from winter frost, another routine threat to crop production. Thanksgiving may bring 80-degree highs, but it's also been known to mark the onset of 28-degree lows, and 3 a.m. in the bone-chilling darkness is not the best time to learn that rats have chewed through important bits of circuitry in the machines protecting your crop.

Wind machine, Churchill-Brenneis Orchard, Ojai.   *John Krist*

The machines—Leavens Ranches has 65 of them—are basically airplane propellers on tall poles, driven by propane-fueled truck engines. By mixing cold, ground-hugging air with warmer layers aloft, wind machines can raise an orchard's average temperature by a few degrees, enough to save the fruit on a frigid night.

And like many other growers, Leavens also is worrying about fallout from a 2006 food-contamination incident, in which at least 199 people were sickened and three people died after eating bacteria-tainted California spinach. The episode prompted calls for new field sanitation and food-safety procedures, whether voluntary or enforced by government, which many farmers fear will increase their production costs and narrow their already slender profit margins.

But autumn is not just a time of worrying. Leavens has stacks of new computer printouts on his desk showing the results of nutrient tests on his orchards. They confirm that the expensive fertilizer applied earlier in the year has made its way into the trees, priming the orchard to produce next year's fruit. And in early December, the packinghouses mail out the checks for the current year's crop.

Those checks are a report card, revealing with unforgiving clarity how well managed—or how fortunate—a farming operation has been over the past year. And it helps establish a goal for the coming months.

"This is the time to regroup and figure out what your games are going to be for next year," Leavens says.

WAITING FOR WEIGHT

Cattle rancher Richard Atmore also is waiting, but not just for rain. In recent weeks his cowboys have been rounding up the herd, tracking down scattered groups of animals on his rugged 6,800-acre spread in the hills north of Ventura so they can be sent to auction. But this year, instead of trucking them to the Tulare County Stockyard in Dinuba, his usual sales outlet, he's been hauling them to a horse stable in Oak View and waiting for corn prices to fall.

High corn prices mean higher costs for the feedlot operators who feed grain to confined cattle for the final 100 days of their lives, marbling the meat with fat and readying them for slaughter. A thousand-pound steer will consume 2,500 pounds of grain during this stage of its life, and it makes more economic sense to ship a half-ton of cow to the grain than to ship more than a ton of grain to the cow. That's why most big feedlots are close to the nation's Midwestern corn belt.

Atmore doesn't sell his animals directly to feedlots; at around 550 to 650 pounds, his free-range steers and heifers will be bought by intermediate operators known as stockers, who will graze the cattle on grass until they gain another couple hundred pounds and are ready to move on to the final stage of beef production.

But when corn prices rise, feedlot operators shave costs by paying stockers less for their animals. In turn, stockers drop their auction bids for Atmore's cattle.

"Right now the price of corn is as high as all get-out," Atmore says. "We decided to hold on for 60 days and see what happens. It won't get worse, and it might get better."

Meanwhile, his cows are hanging out at Ted Robinson Training Stable in Oak View, where horses are schooled in the traditional skills required to work cattle. Some horses are destined for use on working ranches, and others are being groomed for the competition circuit, where Robinson is the winner of multiple championships.

There are a couple dozen horses at the training stable, being prepared for work or for competition. The skills they are being taught, like Atmore's entire operation, are a link to the earliest days of commercial agriculture in California, when Spanish *vaqueros* chased lean, rangy cattle by the thousands across the tawny hillsides.

A STAKE IN STEAK

Here's how the National Reined Cow Horse Association describes such a contest, consisting of three events: herd work, rein work and cow work.

"In the herd work, the horse must cut one steer from a small herd of cattle

Walnut tree, Upper Ojai.

Gary Faye

and keep it from returning to the herd. The horse blocks the movement of the steer in much the same way a defensive basketball player guards the offensive player—crouching low and feinting back and forth. In the rein work the horse maneuvers through a pattern of figure eights, straight runs, lead changes, sliding stops and 360-degree spins. In the cow work—the most challenging of all—the horse must control the movements of a single steer at a dead run, heading it off and turning it both ways along the fence, then bringing it into the center of the arena to circle it once in each direction."

It's fun to watch, and ESPN has televised some of the top competitions. Training young horses to perform these tasks—ideally with minimal guidance from the cowboy—requires a lot of cattle to serve as practice subjects, and that's the role Atmore's animals are playing.

It's a symbiotic arrangement: In return for the free use of Atmore's cows, Robinson provides them free hay. That's worth about $170 a day, Atmore said, and it means his animals will continue to gain weight at no cost to him while he waits for corn prices to drop and beef prices to rise.

Left to forage on dry grass in the sun-cooked hills, the animals probably wouldn't put on any weight at all, he said. At Robinson's place they can pack on about a pound and a half per day.

About 90 of the biggest yearlings in Atmore's herd were rounded up and auctioned off back in July. Atmore figures he'll ship another 100 to 120 animals to the stockyard in Dinuba by year's end.

"This ranch will sell 100,000 to 120,000 pounds of beef (this year)," he says. "And that'll feed a few families."

An organic odor

Jim Churchill's barn smells funny.

It's a relatively new structure, still unfinished in places, and so there are notes of fresh wood in the air, mingling with the vaguely dusty odor of fruit bags and boxes that occupy various corners. Through an open door, the warm November breeze carries the sweet, grassy smell of the surrounding citrus and avocado orchard. But riding over the bucolic aroma is something a bit sharper, vaguely sour and unpleasant.

Turns out it's coming from a pallet of sacks by the big roll-up door. Close inspection reveals them to be filled with smelly brown pellets that look sort of like rabbit food but are a plant fertilizer concocted from ground poultry feathers. Sold by California Organic Fertilizers—which also carries growth-boosters based on seabird guano and bone meal—the feather-based fertilizer is a key element of Churchill's quest to become a federally certified organic grower.

Churchill's 17 acres on the east end of the Ojai Valley are mainly the domain of pixie tangerines, although he and his wife, Lisa Brenneis, also sell other specialty citrus and avocados. Autumn is a transition time for their operation, too; they haven't harvested fruit since early summer, and the first crop of the approaching season—seedless kishu mandarins—likely won't be ready until around Christmas.

That's a couple weeks later than usual, and it's a disappointment, because it means missing out on a market that's been important to the small operation in previous years: pre-holiday gift sales.

"Basically we're vamping, vamping in the orchard," Churchill says.

Vamping means running the irrigation system and hoping for rain. It means whacking and digging up weeds, which under the rules of organic farming cannot be fought using the weed-killer's best friend, a potent chemical called glyphosate. It means starting a new compost heap using wood chips provided by a tree-trimming crew, which will be mixed with manure provided by a local horse owner and left to transform itself into a rich soil amendment.

Or maybe not. As Churchill is discovering, becoming a certified organic grower—which will enable him to sell fruit to retailers specializing in organic products, an important sales channel for small operators—means abiding by a thicket of government regulations. He's still trying to figure out how carefully he must document the care and feeding of the horses that produced the manure, to guarantee that prohibited substances haven't made their way into the compost.

"We've chosen to become organic because that's the only label available to tell people who don't know you what you are," he says. "We think our growing practices meet the health standards of most people who are paying attention."

MORE WORRIES

Churchill figures he could have become a certified organic grower a couple years ago but for a tactical error. He'd already given up petroleum-based fertilizers and chemical pesticides. And he'd begun using municipal yard waste as mulch, to keep down weeds and retain moisture in the soil. But the mulch contained bits of Bermuda grass, which quickly took root and began spreading throughout the orchard.

"It's a ferocious competitor for water and it will strangle trees," Churchill says. He decided that before making the official transition to organic production, a process that requires three years of documented adherence to the federal standards, he'd use glyphosate-based Roundup to kill off the invading grass once and for all.

"I probably should have hired a professional, but I did it myself," he says. "It was stupid, stupid."

He either used the wrong formulation, or the wrong amount, or applied it incorrectly. In any event, he failed to eradicate the tenacious weeds. But by using the chemical, he delayed his qualification for organic certification, which he now expects to occur in time for the 2008 harvest.

Like his counterpart at Leavens Ranches, an operation at the opposite end of the agricultural spectrum in size and type of operation, Churchill worries. New food safety standards developed in response to the 2006 spinach disaster could affect him, too. A proposal to require that cropland be kept free of animal waste, for example, has him wondering how he might persuade coyotes not to defecate among his trees.

And he worries about possible competition on the horizon. Thousands of acres of seedless mandarins have been planted in the Central Valley, potentially flooding the market with cheap fruit. That may weaken consumers' willingness to pay a premium for superficially similar products from small operators, who can't compete on price with mass producers even if they grow a tastier product.

And at least one new variety being developed by university citrus researchers is nearly as flavorful—and ripens at nearly the same time—as Churchill's signature product, the pixie tangerine.

"If it produces well, and it holds up well on the tree, and holds up off the tree, that will be a very strong competitor," he says.

Fall strawberries in bloom, Oxnard Plain.

Gary Faye

Pepper pickers at work, Oxnard Plain.

Gary Faye

CHAPTER EIGHT

Farming on the City's Edge

Autumn is a political season, as well as a time for farmers to worry and prepare for winter. And on Election Day in 1995, voters in Ventura launched a revolution in the way local communities regulate urban development, narrowly approving a controversial ballot initiative requiring a public vote before farmland can be covered with housing tracts, office parks or shopping malls.

Proponents named their campaign SOAR, for Save Our Agricultural Resources, and said it was needed to protect the county's agricultural industry from suburban sprawl. The owners and managers of those agricultural resources, however, were among the most implacable opponents of the ballot initiative.

"We thought the world was going to come to an end," recalls Link Leavens. Much of the 950 acres he and his relatives farm has been in the family for generations, and like many growers, Leavens figured the new law would drive developers out of the area, diminishing the value of his family's most significant asset and forcing him to keep farming even if it was no longer profitable.

So far, however, there's still money to be made growing lemons and avocados. And cropland values have continued to rise, undercutting one argument against the initiative.

"My attitude about this thing has changed," Leavens says. "How can you whine when your net worth is increasing?"

But if the opponents' darkest fears have not materialized as the SOAR era enters its second decade, neither have the fondest dreams of the campaign's supporters. Land-use data gathered over the past 22 years suggest that if the SOAR ordinances have had any effect on the pace of local farmland loss, it has been minor. Each year, Ventura County's cities continue to pave over hundreds of acres of the richest farmland in the world, at a pace nearly equal to that before the laws were adopted.

"SOAR is not about how fast you grow," says Supervisor Steve Bennett, a co-founder of the SOAR movement and a principal architect of its countywide ballot campaigns. "It's about where you grow."

REACTION TO SPRAWL

Farmers were not the only critics when local growth-control advocates launched the first SOAR campaign. Growers were joined by the building industry, which contributed to the unsuccessful opposition campaign and equally unsuccessful post-election legal challenge. The laws, the Building Industry Association of Southern California warned, suffered from "fatal flaws" and were based on erroneous claims that the county was losing farmland at an excessive rate. The BIA also warned that SOAR would severely crimp the local housing supply.

Several years in the making, that first SOAR initiative was a product of public alarm over development patterns in East Ventura, where city officials in the 1980s had approved a series of projects that leapfrogged along the Santa Paula Freeway corridor. Bennett, who was a high school teacher at the time and had not yet run a major political campaign, said Ventura residents turned to the initiative process after they discovered how difficult it was to control land use by electing slow-growth City Council candidates.

The Ventura ordinance, which passed with 52 percent of the vote, requires a public vote before any land designated as agricultural in the city's General Plan can be converted to another use. Following that success, the SOAR campaign expanded and took on a slightly different thrust, as reflected in the new name it acquired and still bears today: Save Open-space and Agricultural Resources.

SOAR initiatives were approved in Thousand Oaks, Camarillo, Oxnard, Simi Valley and countywide in November 1998. Moorpark followed suit in January 1999, and Santa Paula in November 2000. Fillmore's City Council adopted a SOAR-backed growth law in October 2001.

The countywide ordinance is similar to Ventura's, requiring a public vote before land designated as agriculture or open space in the county's master planning document can be converted to urban use. The other cities' laws are different. In general, they establish lines around cities, called city urban restriction boundaries, and require a vote before development can occur outside them. Land inside the lines can be developed without a public vote, even if devoted to farming.

But if SOAR failed to diminish agricultural property values, as Leavens and many of his fellow farmers had feared it would, neither did it halt the paving of farmland, as some voters probably hoped it would.

URBAN EXPANSION CONTINUES

Using high-resolution aerial photos, analysts with the California Department of Conservation's Farmland Mapping and Monitoring Program examine land-use changes every two years in 46 of the state's 58 counties. They calculate the acreage devoted to each of several categories of land use, confirm their calculations by visiting the area, and ask local officials to review any proposed changes to maps and tables. The earliest figures available are for 1984; the latest are for 2004.

Ventura voters approved SOAR in November 1995. That makes 1996 the first year of the SOAR era in Ventura County.

Between 1984 and 1996, Ventura County was losing "important farmland" at a rate of 754 acres a year. After SOAR, the trend reversed and the county gained an average of 591 acres of farmland a year, which would seem clear proof that SOAR worked as intended.

(A catchall designation, "important farmland" includes virtually every type of irrigated and non-irrigated cropland, as long as it was planted and harvested at least once within four years of the mapping date.)

But the picture is more complicated than it seems. Another category of agricultural land—acreage suitable for grazing livestock—was experiencing just the opposite trend. Between 1984 and 1996, the amount of grazing land in the county decreased by 391 acres a year. After SOAR, however, the rate of loss roughly quadrupled, to 1,335 acres a year.

Some of that grazing land was covered in houses, particularly in the east county—where "ranchette" developments of large homes on expansive lots have proliferated—but some of it shifted into the category of "important farmland" as growers plowed, planted and began irrigating hillside and canyon acreage formerly devoted to livestock. That explains how the county could appear to be adding farmland even as cities pushed outward across the cultivated Oxnard Plain and Las Posas Valley.

The change between 2000 and 2002 is a prime illustration. The urbanized area of the county expanded by 2,553 acres in that period, but the amount of important farmland increased by 8,275 acres. Urban expansion into farmland during that time was more than offset by new plantings on former grazing land, which diminished by 9,879 acres in that period.

In fact, important farmland acreage decreased in every census period after the start of the SOAR era except that one anomalous span from 2000 to 2002, when thousands of acres of grazing land were converted to farmland. Take that oddball period out of the calculation, and the average loss was 592 acres a year after the SOAR era began, compared to 754 acres a year before voters began adopting growth controls.

Flowers and fog at Otto & Sons commercial nursery, Fillmore.

Gary Faye

Container-plant nursery, Fillmore.

Gary Faye

Because of the fluidity of agricultural land classification—acreage shifts from one category to another nearly as fast as farmers can uproot trees, fallow fields or install sprinklers—perhaps a better gauge of the SOAR initiatives' effect is the amount of acreage devoted not to crops but to houses and other urban uses. And this is where the SOAR era appears to correlate with a modest decrease in the rate at which Ventura County cities have been consuming surrounding farmland.

Before 1996, the county's urbanized area was growing at an average rate of 1,273 acres a year. Since SOAR, the rate has fallen to 1,120 acres a year—a drop of 12 percent.

"This is an ideal place to live," says Cecil Martinez, who has seen thousands of acres of the Oxnard Plain submerged beneath houses and commercial development since he came to the area in 1972. "Who wouldn't want to live here?"

But even as he understands the impulse that drives landowners and developers to transform fertile ground into homes and businesses, Martinez wonders about the future of his industry as cropland vanishes and new neighbors make it tougher to continue traditional farming activities that involve noise, dust and chemicals.

"How do you balance it out so a farmer can make a living from the land?" he asks.

LAND PRICES RISE

When it comes to agricultural land values—another key issue during the early SOAR campaigns—the county's growth-control laws appear to have had little effect.

Supporters of the initiatives had argued that developers and real-estate speculators, anticipating tremendous profits once they obtained permits for housing tracts and commercial projects, were driving up the value of farmland and pricing growers out of the market. Agricultural landowners indirectly lent credence to that argument by complaining that SOAR would undermine their property values.

But a comparison of agricultural land values before and after the start of the SOAR era reveals no clear indication that the laws have had any effect on the price being paid for farmland in Ventura County.

Those numbers are tracked by the California Chapter of the American Society of Farm Managers and Rural Appraisers, which gathers sales and lease data for crop and grazing land in the state's major agricultural areas. Data are available from the organization's annual reports as far back as 1992.

Between 1992 and 1996, when the SOAR era began in Ventura County, the value of row-crop land increased 50 percent, or 12.5 percent annually. After SOAR, row-crop land continued to appreciate at nearly the same rate, 12.2 percent annually. In 2004, that land sold for up to $65,000 an acre, nearly double what it brought in 1996.

Much of that increase, according to the appraisers' organization, was the result of a bidding war for prime Oxnard Plain land by strawberry growers, who produce a high-value crop that's profitable enough to compensate for high land costs.

Land planted in lemon orchards—still profitable, but with thinner margins and increasing competition abroad—declined in value by an average of 2.75 percent a year between 1992 and 1996. After SOAR, that land increased in value by 1.4 percent a year. Orange orchards lost 5 percent annually in the four years before SOAR; since then, they have increased in value by an average of 14.9 percent a year.

Much of that appreciation occurred between 2002 and 2004, when the selling price of orange land more than doubled. Appraisers say that appreciation was driven by an influx of wholesale nursery operations, which tore out many of the county's unprofitable orange orchards—which were being undercut by lower-cost production in the Central Valley—and replaced them with trees, flowers and other potted plants for sale to landscapers and retail nurseries.

Avocado land, the fourth category tracked by the appraisers' group, was rising in value by an average of 1 percent annually before SOAR. Since 1996 and the adoption of voter land-use controls, the rate of appreciation has accelerated to 4.5 percent a year.

FACING THE FUTURE

SOAR supporters such as Supervisor Bennett believe there's a better way to measure the initiatives' effect on farmland loss than simply looking at changes in total urban and agricultural acreage.

Faced with the challenge of conducting and winning an election before they can build on protected farmland—rather than simply having to persuade a city council majority to approve a zoning change—some builders hoping to profit from Ventura County's pricey housing market have shifted their attention to development of vacant, underused or blighted parcels inside city boundaries. And that, he argues, has saved farmland that would otherwise have been targeted for development.

"The city of Ventura is a classic example," he says. "I don't think there's any question that you wouldn't have had that focus on downtown" without SOAR. And he credits the voter-backed ordinance with a significant policy shift in the city's latest comprehensive plan, which pledges to accommodate growth for the next 25 years mainly within the community's existing urban footprint.

"There's no question that Ventura and Santa Paula would be closer together today if it weren't for SOAR," he says.

Still, he acknowledges that Ventura County farmland continues to disappear, and at a rate that probably means the acreage within SOAR-imposed growth boundaries will be exhausted in some communities long before the laws themselves expire. Although SOAR supporters claimed during the election campaigns that sufficient land remained available inside the urban boundaries to accommodate about 60,000 housing units, that estimate turns out to have been too high.

Two studies, one published in December 2001 by Solimar Research Group and another issued a year later by the Ventura County Planning Division, found that there was enough vacant land inside SOAR boundaries for about 33,000 units, if projects were approved at typical densities. Given the current pace of building, the analysts concluded it is likely that all vacant land inside local growth boundaries will be exhausted long before the SOAR measures expire, which will occur between 2020 and 2030. And that, they suggest, will increase pressure to expand the SOAR boundaries well before the laws come up for renewal.

"Many cities are not husbanding that resource and pacing themselves," Bennett says. "That means you're not valuing land as the precious resource it is."

In the meantime, rising agricultural land prices mean a continuing shift away from the kinds of farming operations many voters probably had in mind when they voted to "save our agricultural resources"—stately citrus orchards, and endless rows of leafy greens—toward higher-value products and production methods that can blur the aesthetic line between farming and urban development: raspberry vines hidden by fabric-covered hoops, trees and shrubs in plastic pots and wooden crates, flowers and hydroponic vegetables grown inside huge glass greenhouses.

"That's not going to make a lot of people very happy," says Earl McPhail, the county's agricultural commissioner.

THE CHEMICAL FRATERNITY

Deliberately maintaining farms and ranches as greenbelt buffers between cities has consequences. Friction over such everyday disturbances as noise, odors, dust and trespassing is common where cropland abuts homes, schools, churches and shopping malls, a common juxtaposition of land uses in Ventura County. But nothing serves as a more potent flashpoint for rural-urban conflict than the use of agricultural chemicals.

It is rare to find documented instances of pesticide drifting off fields and into neighborhoods. But spray rigs and crop dusters provide an anxiety-provoking reminder that the fields and orchards surrounding Ventura County communities are not just a scenic amenity, but a manufacturing zone where the use of hazardous substances is routine.

Leavens Ranches applies chemicals in its lemon and avocado orchards to control pests, boost crop yields and kill weeds. So do the Oxnard Plain strawberry growers overseen by Cecil Martinez. Even rancher Richard Atmore is a member of

Celery field, Oxnard Plain.

Gary Faye

the chemical fraternity, although in a minor way. One of his cowboys often rides the range armed with a machete and a squirt bottle of glyphosate—the active ingredient in Roundup, a popular herbicide that seems to have been named with weed-whacking cowboys in mind—to prevent invasive artichoke thistle from displacing the grasses that feed his range cattle.

The chemical fraternity has many members. According to figures compiled by the Agricultural Commissioner's Office, about 7 million pounds of insecticides, herbicides and fungicides are applied on Ventura County farms and ranches each year.

Farmers are not alone in using chemicals to kill unwanted organisms. The U.S. Environmental Protection Agency estimates that home and garden use accounts for about 20 percent of American pesticide consumption. If that average holds in Ventura County, then local residents use nearly 2 million pounds of dangerous chemicals in their homes and yards every year.

Still, about three-fourths of all pesticide use occurs on farms. And in Ventura County, nearly three-quarters of the farm chemical total is applied to just two crops: strawberries and lemons.

Strawberries are by far the leader at nearly 3.5 million pounds annually, according to data compiled by the Agricultural Commissioner's Office. That's almost half the total, although the crop accounted for just 10 percent of the county's harvested farmland in 2004. Lemons ranked second, at 1.9 million pounds.

Is 7 million pounds of pesticides too much?

The answer is an unequivocal "yes" from anti-pesticide activists and other critics of conventional agriculture, who argue that farm chemical use presents a pervasive threat to human health and the environment.

Outside the ranks of organic producers, which in Ventura County tend to be small operations that serve specialty markets, growers generally argue that chemical use is not a matter of choice but of economic necessity. Many say they would happily kick the chemical habit to avoid the cost, risk, red tape and public opprobrium that go along with it. But they say high production costs and slender profit margins make them dependent on the yield-boosting effects of chemicals to stay in business.

"It's plain old economics," says David Schwabauer. "Part of the way we're able to produce the volume of food that we do is the chemical tools we use."

Some pests that imperil fruit and vegetable production are simply resistant to non-chemical defenses, Schwabauer says. And although weeds can be battled effectively by hand, growers with hundreds of acres find it cheaper to spray than to hire laborers to do the work manually.

The high cost of land is a key factor in the economic equation, growers say. According to the California Chapter of the American Society of Farm Managers and Rural Appraisers, Ventura County row-crop and strawberry acreage is among the most expensive farmland in the state. Coaxing sufficient income from that land to cover the high cost of buying, renting or leasing it often requires intense, year-round crop production.

That level of production stresses soil fertility, allows pathogens to build up in the soil, and leaves little time in a growing season for farmers to respond to diseases and pest infestations without resorting to potent chemical weapons. It also requires that growers plant high-value crops—notably delicate fruits, flowers and vegetables bound for the fresh market—which can be rendered worthless almost overnight if their cosmetic perfection is compromised.

Growers say that as long as the consumer-driven marketplace rewards those who produce unblemished produce and penalizes those who do not, the agricultural industry will have a powerful incentive to continue using hazardous chemicals.

And as long as Ventura County remains a place where suburbs push up against cropland, the potential for chemical-related conflict will persist.

"It's not going to go away," says Rex Laird, chief executive officer of the Farm Bureau of Ventura County.

Like many pesticide critics, Laird believes better communication among farmers, environmentalists and community leaders offers the best hope for reducing rural-urban friction.

"I think dialogue is the answer," Laird says. "Whether we think someone's fears are reasonable is irrelevant."

Washing pimentos, Saticoy Foods packinghouse, Santa Paula.

Gary Faye

Orange orchard, Ojai Valley.

Gary Faye

Epilogue

The year chronicled in these pages began with Jim Churchill leading an impromptu tour of his Ojai citrus and avocado orchard. It was one of those preternaturally warm and bright winter days that, for more than a century, have bedazzled visitors to Southern California.

Twelve months later, he is fighting fatigue, having spent most of the past three nights keeping desperate watch over his orchard as the hardest freeze in nearly two decades wrapped his glossy green trees in a potentially lethal embrace. Punch-drunk tired, he is in the barn packing kishu mandarins in boxes for customers when what he really wants to do is go home and sleep.

Asked how cold it got, he consults penciled entries in a grubby notebook.

"It was 26 degrees at 6 a.m.," he says groggily.

Fruit begins to freeze at 28 degrees.

In a corner of the barn is a crate of recently picked avocados, shriveled and blackening. Over the next few days, every avocado in Churchill's grove ends up looking like that, turned to worthless mush by the elements—a reminder that farmers are in business with Mother Nature, and that Mother Nature is often a lousy business partner.

That's been one of many recurring themes in these pages. Over the course of the year chronicled in this book, growers spoke repeatedly about the challenges they face from pests, weather, global competition, rising land and labor costs, and conflicts with urban neighbors. They discussed how they draw on cutting-edge research and employ 21st century technology to maintain profitability despite such challenges.

But in January 2007, those growers—and the thousands of people who pick, pack, haul or otherwise handle the bounty of Ventura County's fields and orchards—are being reminded of farming's hardest and most enduring truth: A year's work can be undone in a few bitter hours by something as simple and uncontrollable as cold air.

RETHINKING THE BUSINESS

Outside Churchill's barn, although the morning sun shines brightly from a cloudless sky, it is still very cold and jagged icicles hang from the limbs of pixie tangerine trees. Throughout the orchard, crystalline fairy rings of frost mark the locations of tiny sprinklers, and the glazed soil is crunchy underfoot.

Ice-encrusted fruit is an arresting sight, the product of low temperatures and the irrigation system Churchill activated in an effort to save the crop. Heat is released when water freezes—it surrenders the energy it had to absorb to become liquid in the first place—so turning your tangerines into popsicles is paradoxically a way of saving them. Irrigation systems were running all over Ventura County during the cold spell, turning thousands of orchard acres into a vast gallery of surrealistic ice sculptures.

Elsewhere in Churchill's orchard, rows of kishu trees look like giant silkworm cocoons snaking across the stony ground. When they learned about the approaching cold spell, Churchill and his wife, Lisa Brenneis, recruited some friends and together they managed to wrap three acres of kishus in protective blankets of white fabric.

Tangerines wrapped in frost-protection fabric, Churchill-Brenneis Orchard, Ojai.

John Krist

It likely saved the crop. The pixies are probably okay, too, Churchill says, being more cold-resistant than most citrus varieties.

As for the avocados, all 5 acres are probably a total loss. The only real protection Churchill could provide them was to run the propane-powered wind machine, which mixes the cold air at ground level with warmer air aloft and can raise an orchard's temperature a few degrees. Ultimately, that wasn't enough— nor was it for many other local avocado growers—and it did not help that the propane tank ran empty before the cold broke.

Other growers lit oil-burning orchard heaters or smudge pots and fired up their own wind machines. But the effectiveness of those measures dwindles as the

Ice-crusted tangerines, Ojai.

John Krist

cumulative hours of sub-freezing temperatures lengthen, and as the mass of cold air deepens to the point where there's no warm layer aloft to mix with the frigid air at ground level.

"We lit 800 pots last night," Link Leavens says on the sixth straight morning of cold. "And it's not 30-cent oil anymore."

Each orchard heater holds nine gallons of diesel fuel and burns about a gallon an hour, at a cost of around $2.25 a gallon

Exhausted by a week of nearly sleepless nights, Leavens shakes his head over the cost the family operation incurred despite the near futility of some of its efforts..

"I was getting four and a half or five degrees of temperature increase (from wind machines) the night before last." he says. "Last night, it was only one to two degrees."

Berry growers also took a hard hit. The freeze ruined fruit in the fields, and the cold weather prompted the plants to retreat nearly into dormancy. By the time they came back to life, set more fruit and produced a new crop of ripe berries, growers had lost eight weeks of production and missed out on the narrow marketing window when prices and profits are highest. Most of the crop was sold for freezing or processing into jams, preserves and other products.

The freeze followed on the heels of powerful winds, which had already damaged or ruined a substantial percentage of the avocado crop, as well as a fire that swept through lemon and avocado orchards near Moorpark in December 2006, incinerating thousands of acres of trees, including many in the Leavens Ranches orchard. It was a nearly Biblical succession of calamities for some farmers. And the damage was not confined to a single season.

For avocado growers, a truly bad freeze in January doesn't just mean the loss of the maturing fruit already on the trees. It also kills

Fire-blackened lemon tree, Leavens Ranch, Moorpark.

Gary Faye

the branch tips that would be producing flowers in a few weeks—the botanical machinery responsible for manufacturing the next crop—thereby wiping out a second year's production as well.

"We're probably going to rethink the avocado business," Churchill says. His orchard is in a relatively high interior valley, an area only marginally suitable for avocados, which are native to the semi-tropical forests of Central America. The epic freezes of 1990, 1998 and now 2007, combined with the avocado's habit of bearing big crops only every other year even when conditions are perfect, means he's had poor or mediocre seasons more often than not over the past two decades.

LOOKING TO THE FUTURE

Periodically rethinking the business is one of the oldest traditions in Ventura County agriculture, which has seen one crop after another rise in prominence and profitability before declining and then vanishing.

Sometimes it has been in response to particular natural disasters—a pair of droughts and some flooding pretty much ended the hides-and-tallow business of the 19th century rancho period—but more often it has been a response to market forces. Cattle gave way to grain, grain to beans, beans to beets, and beets to nuts, lemons, oranges, and row crops such as vegetables and strawberries. Each transition has been driven by the quest for continued profitability in a changing and very competitive marketplace.

But what if there comes a day when there is no "next crop," when nothing can be grown profitably enough in Ventura County to offset the region's comparatively high and constantly rising land, labor, fuel and regulatory costs? What then? Is it likely that the Ventura County of 20 years from now will look like the Ventura County of today?

If you sit a bunch of growers down in a room and ask them that question, here's the answer you get: Maybe. But it will require cooperation between growers and their urban neighbors. It will require developing new distribution channels to get food from producers to consumers.

It also will require finding a way to properly pay, house and treat the farmworkers whose labor is critical to the industry, without bankrupting their employers. For even in this age of advanced technology, when driving a tractor across a broad field requires being able to operate a computerized guidance system linked to a GPS satellite, food production still comes down to men and women in fields and orchards. It still comes down to callused hands and tired arms and aching backs.

It's a fair bet that almost nobody thinks about this when they grab a plastic clamshell of pre-cut fruit salad from the supermarket shelf. Nor do we typically think about the farmers themselves, or the huge and varied cast of contributors responsible for each and every forkful of food that enters our mouths.

The same could be said, of course, about everything we buy and consume. There's an awful lot of poorly paid and tedious work involved in the manufacture of the computer upon which the manuscript of this book was typed.

But food is different. It is the only product we regularly purchase that literally becomes a part of our living flesh. And so our relationship with it is fundamentally different from the relationship we have with any other consumer product. And we owe it to ourselves to think about how food is produced, the roads it travels from field to plate, and about the implications of that journey for the air and water and the food itself.

At an elemental level, this is the message of *Living Legacy:* Behind every meal is a story.

Many stories, in fact, the stories of all the people whose labor, money, anxiety and hope combine to coax sustenance from soil and sunshine. The stories reach back in time, too, encompassing the parents or grandparents or great-grandparents of those people, who came here seeking work, seeking opportunity, seeking dignity, seeking a chance to put down roots in a remarkable corner of the world.

Understand this, and you will never look at a piece of fruit, a slice of steak or a plate of salad the same way again.

Scorched orchard, Leavens Ranch, Moorpark.

Gary Faye

Hauling peppers from the field, Oxnard Plain.

Gary Faye

References

Aguirre International. *The California Farm Labor Force: Overview and Trends from the National Agricultural Workers Survey.* Burlingame: Aguirre International, 2005.

American Society of Farm Managers and Rural Appraisers. *Trends in Agricultural Land and Lease Values.* Woodbridge: American Society of Farm Managers and Rural Appraisers, California Chapter, 2006.

Blisard, Noel and Hayden Stewart. *Food Spending in American Households, 2003-04.* Washington, D.C.: USDA Economic Research Service, Economic Information Bulletin Number 23, 2007.

Brewer, William H. *Up and Down California in 1860-1864.* Edited by Francis P. Farquhar. Berkeley and Los Angeles: University of California Press, 1966.

California Department of Conservation. *Farmland Mapping and Monitoring Program.* Sacramento: California Department of Conservation, Division of Land Resource Protection (http://www.conservation.ca.gov/DLRP/fmmp/index.htm).

California Department of Water Resources. *California's Groundwater.* Sacramento: Department of Water Resources, Bulletin 118, 2003 Update.

Economic Research Service. *Food Marketing and Price Spreads.* Washington, D.C.: USDA Economic Research Service electronic briefing room (http://www.ers.usda.gov/Briefing/FoodPriceSpreads/).

Falck, Susan. *Evolution of Ventura County Agriculture: 1782-2000.* Unpublished report prepared for the Museum of Ventura County, Ventura, 2005.

Gibson, Robert O. *The Chumash.* New York: Chelsea House Publishers, 1991.

Guilin, Alfonso, Philip Martin and Edward Taylor. "Immigration and the Changing Face of Ventura County." University of California's Giannini Foundation of Agricultural Economics, *Agricultural and Resource Economics Update* July/August 2002.

Hampton, Edward Earl Jr. *Ventura County: Garden of the World.* The Ventura County Historical Society Quarterly 46 (whole year issue, 2002).

Khan, M. Akhtar, Phillip Martin and Phil Hardiman. *California's Farm Labor Markets: A Cross-Sectional Analysis of Employment and Earnings in 1991, 1996, and 2001.* Sacramento: California Employment Development Department, Labor Market Information Division, 2003.

National Agricultural Statistics Service. *2002 Census of Agriculture.* Washington, D.C.: U.S. Department of Agriculture, 2002.

Outland, Charles F. *Man-Made Disaster: The Story of St. Francis Dam.* Revised and enlarged edition. Ventura: Ventura County Museum of History and Art, 2002.

Reed, Jane, Elizabeth Frazão and Rachel Itskowitz. *How Much Do Americans Pay for Fruits and Vegetables?* Washington, D.C. USDA Economic Research Service, Agriculture Information Bulletin Number 790, 2004.

Street, Richard Steven. *Beasts of the Field: A Narrative History of California Farmworkers, 1769-1913.* Stanford: Stanford University Press, 2004.

Teague, Charles Collins. *Fifty Years a Rancher.* Second edition. Los Angeles: Ward Ritchie Press, 1944.

Triem, Judith. *Ventura County: Land of Good Fortune.* Chatsworth: Windsor Publications, 1985.

Vancouver, George. *Vancouver in California, 1792-1794: The Original Account of George Vancouver.* Edited by Marguerite Eyer Wilbur, 3 vols. Los Angeles: Glen Dawson, 1954.

Ventura County Agricultural Commissioner. *Annual Crop Report 2005.* Santa Paula: Ventura County Agricultural Commissioner, 2006.

Packed cartons near the end of the line, Saticoy Lemon Association packinghouse.

Gary Faye

Resources

Ag Futures Alliance
http://agfuturesalliance.net/ventura

Associates Insectary
http://www.associatesinsectary.com

Avocado Commission
http://www.avocado.org

Boskovich Farms
http://www.boskovichfarms.com

Brokaw Nursery
http://www.brokawnursery.com

Calavo
http://www.calavo.com

California Department of Pesticide Regulation
http://www.cdpr.ca.gov

California Institute for Rural Studies
http://www.cirsinc.org/recent.html

California Strawberry Commission
http://www.calstrawberry.com

Center for Comparative Immigration Studies
http://www.ccis-ucsd.org

Center for Immigration Studies
http://www.cis.org/index.cgi

Central Coast Alliance United for A Sustainable Economy
http://www.coastalalliance.com

Churchill Orchard
http://www.tangerineman.com

Community Alliance With Family Farmers
http://www.caff.org

Environmental Working Group
http://www.foodnews.org/reportcard.php

Farm Bureau of Ventura County
http://www.farmbureauvc.com

Food Marketing Institute
http://www.fmi.org

Museum of Ventura County
http://www.venturamuseum.org/library

Limoneira Co.
http://www.limoneira.com/index.htm

National Organic Program
http://www.ams.usda.gov/nop

Natural Resources Conservation Service
http://soils.usda.gov

Pesticide Action Network North America
http://www.panna.org

Pew Hispanic Center
http://pewhispanic.org

Port of Hueneme
http://www.portofhueneme.org

Rural Migration News
http://migration.ucdavis.edu/rmn/index.php

Seminis Seeds
http://www.seminis.com/about

Sunkist
http://www.sunkist.com/home.asp

Sunrise Growers
http://sunrisegrowers.com

U.S. Bureau of Reclamation
http://www.usbr.gov/dataweb/html/ventura.html#general

U.S. Department of Agriculture
http://www.usda.gov

UC Davis Dept. of Agricultural and Resource Economics
http://www.agecon.ucdavis.edu

United Farm Workers
http://www.ufw.org

United Water Conservation District
http://www.unitedwater.org

University of California Agricultural Issues Center
http://aic.ucdavis.edu

University of California Giannini Foundation of Agricultural Economics
http://giannini.ucop.edu

University of California Integrated Pest Management Program
http://axp.ipm.ucdavis.edu

University of California Sustainable Agriculture Research
and Education Program
http://www.sarep.ucdavis.edu

Western Growers Association
http://www.wga.com/public/active/siteBuilder/showPage.php?id=389

Outbuilding, Wheeler Canyon.

Gary Faye

Index

(Page numbers in *italics* indicate photographs)

Roadside scene along Highway 118, Somis.

Gary Faye

Roses at nursery, Fillmore.

Sponsors

PREMIER SPONSORS - $5,000
Limoneira Company   Ventura County Star

MEDALLION SPONSORS - $2,500

| Leavens Ranches, A Family Partnership | Rabobank, NA | Southland Sod Farms | J.K. Thille Ranches | Henry F. Vega, Coastal Harvesting |

BLUE RIBBON SPONSORS - $1,000

AG RX	Calavo Growers	Kipp, Brant, Drummond & Associates	Santa Barbara Bank & Trust
Allied Insurance	Deardorff Family Farms	J.D. McGrath Farms	Terry Farms
Randall & Joann Axell	Faria Family Partnership, Ltd.	Wm. L. Morris Chevrolet	United Water Conservation District
Rancho Abuelos, W.L. Orcutt	Hoffman, Vance & Worthington	Reiter Affiliated Companies	Lorenzo Vega, Rancho Recuerdo

LEADERSHIP SPONSORS - $500

Agromin	Al & Elaine Caveletto	Friends Ranches, Inc.	Mission Produce	Travis Agricultural Construction
R.A. Atmore & Sons	Chase Bros. Dairy	Steven & Barbara Jo Gill	The Nature Conservancy	Underwood Family Farms
Beardsley & Son, Inc.	Churchill-Brenneis Orchard	Grether Farming Company	Pecht Ranch	Fred & Edith Van Wingerden
Elizabeth M. Blanchard Family	Central Coast Farm & Ranch	Ron & Carolyn Hertel	Petty Ranch, Don & Susan Petty	Pyramid Flowers
Arthur Bliss, Farm & Ranch Appraisals	Betsy Blanchard Chess	Bob Jones Ranch	Petty Realty, Chris Sayer & Don Petty	
John W. Borchard, Jr.	DeMartini Enterprises Paul & Ann DeMartini	Index Fresh	The William L. Reiman Family	Ventura County Coastal Association of Realtors®
Hank & Ellen Brokaw	Farmer's Irrigation	Gene & Sally Mabry	William G. Scholle Ranch	Ventura County Economic Development Association
Julie Bulla	Ferguson, Case, Orr, Paterson LLP	Tom & Brianne McGrath	Tosh Shinden, Sunnyland Nursery	
California Strawberry Commission	Finch Family Farms	John & Carolyn Menne	Somis Pacific & Oxnard Citrus Sam & Carol McIntyre	Nancy Worthington in Memory of William E. Worthington

PARTICIPATING SPONSORS - $100

| Terry & Kathy Bergeron | Claberg Ranch, LLC | Oxnard Lemon Company |
| Boskovich Farms | Logan & Carol Hardison | Richard & Gail Pidduck, Santa Paula Creek Ranch |

Strawberries in bloom near Camarillo.

Gary Faye

Squash display at Underwood farm, Tierra Rejada Valley.

Gary Faye

Cabbage field, Oxnard Plain.

Gary Faye

Special Thanks

FARM BUREAU OF VENTURA COUNTY

Officers

Henry Vega, *president*

Scott Deardorff, *1st vice president*

Leslie Leavens-Crowe, *2nd vice president*

Josh Pinkerton, *secretary*

Will Pidduck, *treasurer*

Directors

David Borchard

Gus Gunderson

Ed McFadden

David Murray

Steve Onstot

David Schwabauer

Edgar Terry

Fred Van Wingerden

Staff

Rex Laird, *chief executive officer*

UNIVERSITY OF CALIFORNIA HANSEN TRUST

Advisory Board Members

Christine Casey

Sue Chadwick

Scott Deardorff

Ben Faber

Christopher Mann

Don Rodrigues

Chris Sayer

Sandi Shackelford

Edgar Terry

Staff

Larry Yee, *county director, UC Cooperative Extension*

Patricia Verdugo Johnson, *business services administrator*

BOOK PROJECT COMMITTEE

John Broome Jr., *chairman*

Betsy Chess

Marty Fujita

Sheri Klittich

Tim Schiffer

Carole Topalian

Winter scene, Upper Ojai.

Gary Faye